Introduction to Plant Ecology

Introduction to
PLANT ECOLOGY

A Guide for Beginners in the Study of Plant Communities

by A. J. Willis

A completely revised edition of the late Sir Arthur Tansley's book of the same title

Distributed in the United States by
CRANE, RUSSAK & COMPANY, INC.
347 Madison Avenue
New York, New York 10017

London · George Allen & Unwin Ltd
Ruskin House   Museum Street

Printed by offset in Great Britain
in 11/12 point Times New Roman type by
Alden & Mowbray Ltd
at the Alden Press, Oxford

Preface

This book had its origin in Tansley's *Practical Plant Ecology*, first published in 1923; new impressions of this text were issued in 1926 and 1932. Subsequently the scope of the book was widened, the balance being shifted in the direction of less emphasis on practical aspects, so as to give a more general coverage of the subject. In 1946 a revised and enlarged second edition was published under the title of *Introduction to Plant Ecology*, to be followed in 1954 by a third edition but with only minor revision and additions.

In the preparation of the present text stemming from Sir Arthur Tansley's book it was felt desirable to make extensive change and to incorporate many additions, in the light of great advances in ecology in recent decades in order to provide a broad and updated treatment. Nevertheless, an attempt has been made to retain the character, style and scope of earlier editions and also much of the general approach of Sir Arthur Tansley to ecology. The lay-out of the book has been only little altered, the main difference being the addition of three new chapters referring to some of the major growth points of the subject. These chapters concern quantitative procedures, mineral nutrition and features of the ecosytem, including nutrient cycling, energy flow and productivity.

Developments in ecology have necessitated change in virtually every chapter of the book, so that relatively little of the original now remains, especially in Part IV (Habitat) which is much enlarged. However, the section on the structure, distribution and development of vegetation is less drastically altered, and the system of classification of plant communities advocated by Sir Arthur Tansley is retained although mention is made of other ways in which vegetation may be classified. Many of the illustrative examples used in the early editions of the book are still as apt and valid as they ever were. Accordingly a number of these have been retained, although supplemented by examples selected from more recent studies.

A new feature of the present edition is the inclusion of a series of plates, illustrating a range of types of vegetation and various features of ecological interest. Most of the Figures of former editions have been eliminated and others added, notably in the chapters on quantitative methods and on soil.

In this extensive revision I have received considerable collaboration from many friends and colleagues who have assisted in numerous ways. In particular I am indebted to those who have given special help in the preparation of drafts of certain chapters. In this respect I am especially grateful to Dr B. Hopkins (Chapter 11), Dr P. S. Lloyd (Chapters 14, 18), Dr M. H. Martin (Chapters 15, 17), Mr L. F. H. Merton (Chapter 18) and Dr I. H. Rorison (Chapter 16). Among the many who have provided information, made valuable suggestions or assisted in other ways, special thanks are due to Miss A. E. Barling (for drawing Fig. 10.4), Mr L. F. Curtis, Dr G. C. Evans, Dr P. J. Grime, Professor J. L. Monteith, Dr M. C. F. Proctor and Dr D. J. Read. To Dr Proctor too I am much indebted for the generous supply of first-class photographs, a number of which are incorporated here. Acknowledgment of the source of these and of other photographs, as well as of the Figures, is made beneath the individual items. To my publishers I am very grateful for their patience, care and efficiency in the production of this book.

A. J. WILLIS

Department of Botany
University of Sheffield
April 1973

Contents

Plates

Figures

Part I

INTRODUCTORY

Chapter 1

What is Ecology?

The word ECOLOGY is derived, like the common word *economy*, from the Greek οἶκος (*oikos*), meaning *house, abode* or *dwelling*. In its widest sense, ecology is the study of plants and animals *as they exist in their natural homes*; or better, perhaps, the study of their *household affairs*, which is actually a secondary meaning of the Greek word. Interpreted broadly, ecology embraces the study of all aspects of living organisms with respect to their existence in nature; but in particular it is concerned with the relationships of plants and animals to one another and to their surroundings or environment.

In this book attention is confined to plant ecology, which is, for several reasons, more readily accessible to the beginner than animal ecology. Much of the latter is not easily understood without some considerable knowledge of the vegetation, because very many animals depend directly upon plants for shelter, while all depend upon them, directly or indirectly, for food. Plants form the basis of all life as it is lived upon the earth because they alone (apart from some bacteria) have the power of producing organic matter from inorganic, of building up living substance from the simple raw materials carbon dioxide, water and mineral salts. Animals can only use the products of this synthetic power of plants, either directly by eating them (herbivorous animals), or by eating other animals which have fed upon plants. In a favourable climate and soil, plants cover the ground more or less completely, thus forming a natural framework, or basis, for the study of the living populations of the globe. In this way they determine not only the food but largely also the shelter and general conditions of life of animal communities and, to some extent, even that of man, although of course man has the power to exercise considerable control over his environment.

As implied above, ecology must include the study of the 'household affairs' of animals, including man, not only because animals form an important part of the life existing on the surface of the earth, but because the effects of animals upon plants are numerous and far-reaching; while man occupies a unique position owing to his far-extended control over nature. Consequently anything like a complete study of the ecology of a plant community necessarily includes a study of the animals living in or feeding upon it. The influence of man upon plant communities is of first importance in all but the uninhabited and the most sparsely inhabited regions of the earth. As we shall see in later chapters, in ecological interpretation we can never afford to lose sight of past and present human activities in their effects on the vegetation of countries which have been long inhabited and densely populated, like those of western and central Europe. But though we must thus constantly take account of the effects of animals upon plants, we shall here be concerned entirely with plant ecology—our centre of interest will be the plants themselves.

It is clear that in the wide sense defined above, plant ecology cannot properly be considered a separate and distinct branch of botany, since it must include a great number of topics which belong to older, well-recognized divisions of our knowledge of plants. Thus, if we are going to study the 'household' affairs of plants as they grow in nature, we must first of all learn to recognize the features characteristic of specific groups of organisms, and more especially to distinguish the different kinds or species of plants with which we have to deal; for this purpose we must have some knowledge of *taxonomy* and *classification*. To this we must add knowledge of *genetics*, the science of heredity and variation, on which depend the origin and maintenance of existing 'taxonomic units'. Then we must understand the construction of the plant body, the differences between its different parts, how they develop from the seed or from one another—and this is a part of *morphology*. Further, we must know something of the cellular construction or minute *anatomy* of plants if we desire to penetrate at all deeply into the reactions of plants to the different environments in which they grow. Again, we must study how far plants depend on insects, the wind or other agencies for pollination and hence fertilization, or how far they pollinate themselves, the ways in which they spread from place to place, the means by which they propagate themselves and are dispersed (such as by fruits, seeds, rhizomes, and runners). All these last-mentioned topics may be referred to under 'biology', but this term more strictly relates to the science of life as a whole. Finally, every attempt to determine the actual causes that underlie the ability of some kinds of plants to flourish in particular situations while others cannot, will certainly lead, not only to questions of the means of dispersal already mentioned and of the influence of animals and of human activity, but also to a study of soil (*pedology*), climate (*climatology*) and particularly the local climate directly affecting individual plants, the microclimate, in relation to different species. These last investigations lead to a study of the physical and chemical relations of the plant to its habitat, involving some of the most difficult problems of *plant physiology*.

The field of *physiological ecology*, which has developed extensively in the last two decades, has led to the understanding of much of the ecological behaviour of plants in terms of their physiology. In such studies attention has been directed, for example, towards the water relations of plants in respect of water availability in habitats and to the requirement and uptake of mineral nutrients in relation to the nutrient status of soils. But many physio-ecological problems remain to be tackled.

Another important aspect in which progress has been rapid in recent years is that of the *dynamics* of ecological systems. Many changes occur in the integrated community of the organisms and their controlling environment—the *ecosystem* as it is often called—some of them of a cyclical kind, and fundamental to the continued existence of the community. The cycling of essential nutrients, involving their uptake and use by growing plants, and the return of the nutrients to the soil on death and decomposition and their subsequent re-utilization, as well as the flow of energy in ecosystems, are basic considerations of this kind of first-rate importance.

The development of all aspects of ecology has been aided by the increasing use of mathematical treatment of results obtained in the field and laboratory and by substantial advances made in quantitative methods. Statistical procedures are especially valuable in giving precision to ecological observations, such as those concerning the relationships between plants and their environment, and allow an objective analysis of plant distribution and community structure. Such an approach is a valuable complement to qualitative, descriptive studies.

Thus it becomes clear that plant ecology in the wide sense is more a means of approach to a large part of detailed botanical study than a name for a special branch of the subject, such, for instance, as *cytology* (the study of cells), *palynology* (the study of pollen), or, again, the study of a particular group of plants such as the mosses or the fungi. Alternatively, it may be regarded as a synthesis and integration of the special knowledge obtained from many different aspects of botany in relation to the life of plants in their natural habitats. The appeal of ecology depends partly on its broadness and the way in which it bears upon a wide range of aspects, serving to link them together. The continued contact with plants in their natural habitats constantly brings up many and diverse points of interest, and much can be learnt by careful observation in the field, especially if this can be spread over a considerable period.

Whereas ecology must be studied primarily in the field, however, observations made during the course of field work may raise problems which can be profitably studied only by experimental methods in an environment more or less rigorously controlled. In garden plots, plants can be observed in the absence of competition, or their performance in particular soils assessed; in a glasshouse or in growth rooms it is possible to study them under climatically controlled conditions also.

A parallel may be found in the study of man. The human anatomist and the human physiologist have, each in his own sphere, a profound knowledge of man, and the two together can give a fairly complete general account of the structure and working of the human body. But no one would contend that such knowledge covers the field of what we may know about man and his activities. It is not sufficient to study the structure of his dead body in the dissecting room, or the functions of his organs and tissues in the physiological laboratory. To learn what man actually is and does in the world we have to go out into the world and study him as he lives and works among his fellows. And the same is true of plants.

Plants are gregarious beings because they are mostly fixed in the soil and propagate themselves largely in social masses, either from broadcast seed (or spores) or vegetatively, by means of rhizomes, runners, corms or bulbs and sometimes by new shoots ('suckers') arising from the roots. In this way they produce *vegetation*, as plant growth in the mass is conveniently called, and in general this is found to fall naturally into *plant communities*, or units of vegetation, although the boundaries between these units cannot always be sharply defined. Now these plant communities have structures, activities and laws of their own. Each has an internal economy depending on the relations of its individual members to one another; each also has an origin, history and fate. Particular communities can exist in some places and not in others, depending on the conditions of soil and climate and on their relations to other plant communities and to animals. Within the larger communities smaller ones exist. In these features we recognize parallels with the nations, tribes, and societies of mankind, though the members of plant communities are not so closely knit as the members of human, and even of the higher animal, communities, by a complex physical and psychological interdependence. Plant communities are also essentially different from human communities in that they are commonly composed not of a single species of organism, but of several or many different species living together.

The main causes of the specific structure and individuality of a given plant community are: firstly, the fact that only those species can be present in it which exist in the particular part of the world, and which are able to reach the particular spot; secondly, that only those can be present which are able to exist under the given conditions of life and in competition with the other species present; and thirdly, that in many communities certain species can survive only in the presence of others and the environment they create. For instance, the occurrence of some of the 'shade plants' on a forest floor is dependent on the trees which cast the shade, and in desert conditions some small annuals and perennials are found only in the immediate vicinity of larger plants around which wind-borne particles are deposited and accumulate into mounds.

The detailed investigation of plant communities, and especially the application of quantitative methods and statistical procedures in the analysis of their structure and of the distribution of individual species within the community (whether, for example, the plants are clumped together or are randomly dispersed), is essentially a recent study, although types of vegetation and their dependence on conditions of life have long been recognized. The active modern study of types of vegetation began at the turn of the century, being much influenced by the pioneering work of Professor Warming of Copenhagen published in the German edition of *Ecological Plant Geography* in 1896. This directed interest to the study of plant communities, with which 'ecology' tended to become identified in Great Britain. But this identification is not justified, ecology in a wide sense, as we have seen, being much broader. *Synecology*, from the Greek σύν, *together*, is now often used to refer to the study of communities, as distinct from *autecology* (Greek αὐτος, *self, by oneself, alone*) for the study of the ecology of individual species, and many important autecological investigations have been made in recent years. The study of plant communities as such is now frequently referred to as plant sociology or phytosociology, especially by Continental and American ecologists. By them the meaning of the word ecology may be confined to the study of the *habitat*, the *oikos* itself, of a plant or community, i.e. of the sum total of the effective conditions under which the plant or community lives in a given spot. This is certainly a strictly logical use. Nevertheless, in this book the word is employed in the broader sense described earlier, for it is important to keep the emphasis on ecology as the approach to botany through the direct study of plants in their natural habitats. In this approach, a knowledge of plant communities, their structure, economy, origin and fate (plant sociology) must bulk very large.

Ecology or environmental biology can be seen to embrace many disciplines. The modern ecologist must be able to observe, to experiment, to interpret and to integrate information from diverse sources. The earlier ecological studies, where the emphasis was strongly on description, are now being augmented by investigations of an experimental kind in which functional aspects of the plant are being increasingly explored. In this way we are gaining a more complete understanding of the structure and functioning of ecological systems, be they as small as a tiny pool of water or as large and complex as a vast tract of tropical forest.

Chapter 2

Natural and Semi-natural Vegetation

By natural vegetation we mean of course vegetation primarily due to 'nature' rather than to man. If we take extreme cases, a virgin forest is clearly natural, while a wheat or root crop is clearly not. But we have to recognize at once that there are a great many cases intermediate between these two extremes. If we leave out of account all the genuinely virgin, untouched communities on the one side, and all sown field crops and plantations on the other, we find that large parts of the vegetation of a country like Great Britain, more especially of the north and west, but considerable tracts also in the south and east, owe their character partly to nature and partly to human activity. If the vegetation itself is spontaneous, i.e. has occupied the ground without the aid of direct human action, but has nevertheless been partly determined or markedly modified by man or his animals, we class it as 'semi-natural'.

Thus natural woods which are 'selectively' felled (i.e. from which single trees are periodically taken out), but not 'clear felled' and replanted, and those which are regularly coppiced are semi-natural. Heaths and moors which are periodically burned or regularly pastured, grassland which has not been sown but which is regularly pastured or mowed, and marshes which are drained and pastured or periodically cut, provide further examples of semi-natural vegetation. The great bulk of the forest and 'waste land' of the British Isles is in this condition. True 'virgin' communities of any size are rare, indeed, almost absent, except on the sea-coast and in the remoter mountain regions. But there are a good many which are substantially natural, having been altered by man only by occasional felling, pasturing or burning, and many more which are semi-natural, i.e. which represent a definite modification of a natural community, and are kept in their existing condition only by the activity of man.

The degree to which man has influenced an apparently natural community varies, of course, very considerably, and has often to be made the subject of special investigation. Continued selective felling of the trees in a natural wood, and especially the constant cutting out of certain kinds only, will, for instance, gradually alter the proportional composition of the wood, and sometimes its whole character. Again, the opening of the wood canopy and the consequent letting in of light will kill certain woodland plants and promote the growth of others. It will also allow the entrance of herbs and grasses which could not grow at all in the deep shade, and these will tend

20

to suppress those true shade plants of the woodland floor which have survived, and to compete with the species whose growth has been stimulated, so that the constitution of the ground vegetation may be entirely altered. Pasturing and burning of grassland or heath will destroy some species and severely check the development of others, while certain species will shoot again quickly after being eaten down or burned, thus altering the composition of the herbage. Others, again, which could not establish in the original plant community, will invade burned or very heavily pastured land, because they find there open or sparsely covered spots which would not exist apart from the burning or overgrazing. Lowering the water level of a marsh or fen, or of low-lying alluvial land, by drainage, will kill out certain species and weaken others, while it will admit fresh colonists that flourish in a drier soil. All these effects, and others of a similar nature, have constantly to be taken into account in studying semi-natural vegetation.

The view is sometimes held that only perfectly 'natural' vegetation is a proper subject for ecological study. If this were true, the ecological field in Great Britain would be very limited indeed. Fortunately, the belief is entirely mistaken. The laws governing the behaviour of plants, their relations to one another, and their formation of communities, are the same whether the activity of man and his domestic animals plays a part or not. It is true that ecological problems are complicated by man's activity—fresh factors have to be considered. But the plants themselves are behaving in the same way, tending towards the same effects, whether man is at work or not. The only practical difference is that the plant communities and their distribution (apart altogether from actually cultivated areas) are to a considerable extent changed in countries where most of the land has long been subject to human interference from those still almost untouched or only little affected by human agency.

In the countries which have long been the seats of civilization it is not always a simple matter to reconstruct the 'original' vegetation. But this can usually be done with a fair degree of certainty when sufficient detailed knowledge of the behaviour of the native plants, and the communities which they form, has been obtained and a comparison made with neighbouring countries having a similar climate and flora, but perhaps with different methods of forestry and agriculture. On the other hand, man is always unwittingly performing ecological experiments on a small or a large scale, experiments which the ecologist can watch and the results of which he can trace out and record, thus slowly gaining an extensive knowledge of the capacities of plants, and of their reactions to changed conditions, which the observer in a 'virgin country' cannot easily acquire.

It is true that the farmer, landowner, or 'local authority' sometimes carries his experiments further than the ecologist would desire. The observation of a process of recolonization of a piece of cleared land, for instance, may be rudely interrupted by digging for gravel, road-making or building. Regrettably, interesting and instructive stands of vegetation are often destroyed by man's activities: peat may be extensively removed from peat bogs; marshes may be drained; vegetation may be covered by spoil from pits; major excavations and infillings may be undertaken in the construction of motorways. But the careful observer can learn a great deal by taking advantage of these opportunities to study changes following these alterations in the environment. For example, when a marsh is drained, some of the plants—those dependent on a high water table—disappear, whereas others, characteristic of drier habitats, may be seen to invade the area.

Even the most highly cultivated parts of the country, where the land is almost all under the plough, offer numerous opportunities for ecological study. Such are great tracts of the eastern counties of England, considerable parts of the southern counties, parts of the Midlands, and the south Lancashire and Cheshire plains. First of all there are the roadsides and hedgebanks, which bear semi-natural plant communities—the former, when they are in the form of 'grass verges', often very similar to those of pastureland or on barren sands of heathland, the latter consisting mainly of 'marginal' woodland species, i.e. those which grow on the edges of woods under similar conditions of soil and climate. These both provide good subjects for study and comparison. Secondly there are the weeds of the arable land itself, which differ according to climate, soil and crop rotation. And finally there are the crops themselves, which are by no means the least interesting communities, though they are entirely artificial. Scientific agriculture, indeed, is largely applied ecology. The field crops which *can* be successfully cultivated on a given piece of land depend, of course, on the climate and soil, just as does the natural vegetation; but the crops which actually *are* cultivated depend also upon economic considerations of effective demand and a profitable market, in so far as they are not used locally.

The grass fields of 'permanent pasture' which occupy so much of the Midlands and West of England (though considerably less than at the turn of the century) are really semi-natural plant communities, for though they may in many cases have been 'laid down to grass' by sowing ploughed land with seed, they quickly become modified by the natural immigration of herbs and grasses, so that sometimes few of the kinds of grass sown ultimately remain. The composition of the flora of permanent pasture varies with the soil, the water supply, the proximity of neighbouring natural vegetation, and also with the kind and amount of pasturing and manuring. The grazing regime is always an important factor.

Plantations of native trees on arable or grassland or on old cleared woodland may develop into communities almost indistinguishable from natural woodland. This is where the same species are used that naturally form woods on the same soil. Woodland plants colonize the plantation, which gradually approximates to the structure and composition of a natural wood—all the quicker, of course, if it was planted on an old woodland site. Woodruffe-Peacock (1918) described such an oakwood on clay soil more than a century after it was planted on arable land, and it had acquired the character of a native wood in nearly every respect. A large number of spruces had been planted with the oaks, and while some of these were still present, many had gone and none were flourishing: eventually all would surely disappear. Larch failed equally. Ash, wych elm and sycamore[1] were the only trees planted with the oaks that really did well, and these are species that one might find in a natural British oakwood.

Plantations of exotic trees, particularly conifers—spruce, pine and larch are the kinds most commonly planted in this country—provide examples of artificial communities, differing, as regards the ground vegetation, more or less considerably from our natural woods. Sometimes they are invaded by 'weeds', sometimes certain woodland plants obtain a footing in the plantation, and sometimes the soil remains or becomes virtually bare. In any case the plantation affords some opportunity for ecological observation and comparison, even where the results are chiefly negative.

Thus there is plenty of work for the ecologist in highly cultivated districts, though it is not quite of the same kind as in regions which are mainly occupied by natural or semi-natural vegetation. Even building sites, railways and roads which have been laid out and left derelict for a time, and similar places, often present the material for interesting study. Such habitats are not of course natural, either in origin or character. The soil is peculiar—indeed, it is not 'soil' in the narrower sense—because it contains little or no *humus* (derived from the decayed remains of plants); it can be colonized only by a certain selection of species—mostly, though not wholly, the weeds of roadsides or of ploughland. But these species and their powers of colonization and relative persistence furnish ecological problems of considerable interest. Broadly they behave in much the same way as the first colonists of 'natural' dry areas in which there is little or no humus. Where there is rich soil containing much combined nitrogen and other nutrients derived from organic refuse, the ground will be occupied by special kinds of plants, of which the common stinging nettle is a good example. Such plants are often specially luxuriant about farmyards, the lairs of cattle and similar places, and have been referred to as *nitrophilous*, but studies by Pigott and Taylor (1964) have shown that at least for the stinging nettle the enhanced supply of available phosphorus is more important than the high level of nitrogen.

Even when a habitat owes its existence directly to human activity, it may be occupied by a perfectly natural plant community, i.e. a collection of

[1] Sycamore is not a native tree but has established itself all over the country since its introduction in the fifteenth or sixteenth century and behaves just like an indigenous tree.

plants whose composition owes nothing to human agency, and might equally well occur in a 'natural' habitat. Thus the stones of a wall built of limestone blocks will be colonized by just the same lichens and mosses that cover the surface of natural limestone rocks, and the crevices by the same flowering plants and ferns that we find in the fissures of the same rocks. A similar situation exists with a wall of sandstone, or dolerite, or granite, the flora to some degree reflecting the neighbouring vegetation (Holland 1972). An artificial pond will be invaded by much the same plant communities arranged in much the same way as a natural lakelet on the same soil and in the same surroundings, a canal by the same communities as a sluggish river, and so on.

From all of these observations we draw the very obvious conclusion that it is not the activity of man at large that is significant to the plant ecologist, but the actual conditions which he creates. If human activity destroys a large number of plant communities and plant habitats, and modifies, to a greater or lesser extent, many more, it also produces fresh habitats and fresh plant communities, and thus provides fresh opportunities for ecological study on every hand.

Part II

STRUCTURE, DISTRIBUTION AND DEVELOPMENT OF VEGETATION

Chapter 3

The Units of Vegetation (Plant Communities)

Since plants are, for the most part, gregarious in their occurrence, we can never get any deep insight into their ecology unless we consider them as members of the communities in which they naturally grow. The 'ideal' method of study might be to investigate each species separately, until we knew in detail its life history, the methods by which and the rate at which it could spread, and its behaviour under different conditions of climate and soil; only when we had obtained this information would we proceed to study the species as it existed in communities along with other species.

This ideal method, however, leading to such a complete knowledge of the ecology of any one species, means many years of observation and experiment entirely devoted to that one species. Nevertheless, it is very desirable that detailed autecological investigations of this kind should be undertaken. Such thorough studies as have so far been made, as for example on the bluebell (*Endymion non-scriptus*) (Blackman and Rutter 1946–50), the great fen sedge (*Cladium mariscus*) (Conway 1936–8), and the alder (*Alnus glutinosa*) (McVean 1953, 1955–9), are of very considerable value in leading to a fuller understanding of the ecology of individual species, but much more intensive work of this sort is required. Notable progress in this direction has, however, already been achieved in respect of a substantial number of species, broadly based ecological accounts of which are given, from 1941 onwards, in the *Journal of Ecology* under the *Biological Flora of the British Isles*. These accounts include information about, for example, the morphology, reproduction, dispersal and distribution of species, their habitats and the plants with which they are associated, and their behaviour in relation to soil conditions and climate. It is planned that such studies will ultimately extend to all British species; even for some of those species for which accounts are available there is still much to learn. Some suggestions as to the lines on which investigations of this kind can be best pursued are given in Part V.

There are, however, many facts which we can find out about the plant communities before we possess an exhaustive knowledge of the autecology of the individual species which compose them. Indeed, the study of a community is one of the best ways of suggesting the most important problems presented by the particular species which make up the community. The study of a plant community always and necessarily drives us back to the individual species, and in this way our interest in purely autecological problems is most likely to be aroused and sustained. No apology, therefore, is needed for beginning with *synecology*. Here, of course, a first essential is to be able to distinguish and name accurately the species of which the communities to be studied are composed.

Some Examples of Plant Communities

If we take a walk through the English countryside and look at the vegetation around us, we shall be able to distinguish without difficulty the purely artificial or culture communities, such as cereal, root, mustard or clover crops, and plantations of larch, spruce or other trees, from the natural and semi-natural vegetation of various kinds. The exact degree to which these latter have been determined or affected by man may be a matter of more difficulty.

The meadow through which our footpath leads (permanent pasture or hay crop) is obviously mainly so determined, though the plants may all be genuine natives. Here we have an example of a plant community determined by man but dominated by different species of native grasses.

The copse which we presently pass has a different status. It may have been planted, but it may be a modified remnant of primitive woodland. The dominant trees and shrubs are very likely those which would have been there in any case, though their form and perhaps their relative proportions have been determined by felling and coppicing. The copse, then, is quite possibly a modified example of the natural forest community, dominated very probably by oak and hazel,

Presently we come out on to a stretch of heath dominated by the common ling (*Calluna vulgaris*). This is very likely determined by a difference of soil corresponding with a distinct geological formation, for instance a sand. It may be kept in its condition as heathland by pasturing or burning, or both, and this may be evidenced by the birches, in many places pines, and sometimes oaks, which grow on its edges, and by the patches of scrub or single bushes dotted about it. These may represent the tendency of woody vegetation to re-establish itself on the ground; such re-establishment might soon be successful if it were not for the constant attacks of grazing animals.

On the other hand, the soil may be unsuitable for tree growth, as shown by the yellowing and dying of young oak seedlings. The heath, then, is certainly a natural or semi-natural community, quite distinct from the meadow and the copse. It may be a stage in the development of forest, or it may represent the only type of vegetation that the particular soil and climate could naturally produce.

In wet hollows of the heath there will be collections of species different from those of its general surface, for instance the cross-leaved heath (*Erica tetralix*), the purple moor grass (*Molinia caerulea*), and perhaps bog moss (*Sphagnum*), butterwort (*Pinguicula*) and sundew (*Drosera*). Here is another distinct community, evidently determined by the wet soil, and also, as we should learn by further comparison, by the very acidic soil water.

Beyond the heath area we will suppose the ground slopes down to a riverside, bordered by a belt of fen or marsh, liable to be flooded by the river after heavy rains. The fen or marsh represents yet another plant community, covered with species of grass and sedge, among which may be many other plants which grow only in wet places, but different from the plants of the wet hollows in the heath. The soil of the fen or marsh may contain about the same amount of water as that of the wet heath, but it is not acidic, and supports a quite different vegetation.

The edge of the river itself may be lined by a reedswamp, composed of tall grasses, sedges or bulrushes. In the water are other plants, with floating leaves if the stream is sluggish; and others again wholly submerged. All of these last-mentioned communities are essentially natural, though they may be modified by human activity in various ways.

It is very easy to see that the various units of vegetation are not all of equal rank. Thus within an oakwood there may be a localized belt of ashes or alders following the banks of a stream, or a patch of bluebells in one place and not in another; on a heath there may be local patches of a species of grass or of moss. And clearly the lichens covering the bark of the trees in a wood cannot be considered as a community of equivalent status to the woodland itself. Yet all these are units of vegetation. We apply the term *plant community* to any such unit of whatever size or rank. Thus the deciduous forest of western and central Europe is a plant community, and so are the submerged water plants in a pond, or the green coating, often largely composed of the minute alga *Pleurococcus*, on a damp wall or tree-trunk. *A plant community is any collection of plants growing together which has a certain unity and individuality.* A particular assemblage may occur repeatedly while closely comparable ones can be recognized in similar habitats.

THE PLANT FORMATION

The largest and most comprehensive kind of plant community is the *plant formation*. Plant formations correspond in a general way with those plant communities which are recognized in common language as fundamentally distinct *types of vegetation*.

First of all we have the great *climatic vegetation types* of the world, determined by the well-marked types of climate prevailing over wide areas of the earth's surface. Some of the most conspicuous and widespread of these are the evergreen tropical rain forests of the Indo-Malayan region, of parts of equatorial Africa and of central America; the deciduous 'summer' forest of western and central Europe, the eastern states of North America and of parts of eastern Asia, and its evergreen counterpart in south-eastern Australia; the northern and subalpine forests of needle-leaved coniferous trees; the temperate climatic grasslands (prairies and plains) of the United States and Canada and the steppes of southern Russia; the deserts of North Africa and south-western Asia, of parts of South Africa, of Chile and of parts of western North America. Each of these regions is characterized by a specific kind of vegetation, 'formed', in a very real sense, by the prevailing climate, and known as a *climatic* plant formation. The European deciduous summer forest formation, to which our own native deciduous woodlands belong, has the same general character as that of eastern North America because it exists in the same general type of climate, and both formations belong to the same *formation-type*. Similarly, the tropical rain forest formation of Indo-Malaya has the same general character as that of west Africa and of central America and belongs to the same formation-type, though each of these formations has its own peculiarities.

Besides the great climatic formation-types there are others determined, not *primarily* by climate, but by conditions such as the general nature of the soil which they occupy. A good example is the *reedswamp type* which grows in shallow water on the edges of lakes and slow rivers all over the world, vegetating the soft waterlogged mud or silt and in some places (e.g. the delta of the Danube), where this kind of soil is very extensive, covering vast tracts of country. Such a formation-type is largely independent of climate, though the reedswamp formations of different parts of the world differ in the particular species of which they consist. Other formation-types, determined in a similar way and more or less independently of climate, are the sand dune and salt marsh formation-types, which are also primarily determined by the soil and other local conditions in which they grow. All these formations, determined primarily by soil, are called *edaphic forma-tions* (Greek ἔδαφος, *edaphos*, the soil, ground) in contrast with the climatic.

There are, however, some formations, such as the *heath formation*, which are determined partly by soil and partly by climate, i.e. they can exist only within rather a narrow range of climatic variation and in suitable soils. This shows that plant formations cannot always be rigorously divided into climatic and edaphic; but appropriate climatic conditions naturally have to be present.

Formation-types can also be recognized in semi-natural vegetation. Such a formation-type is the pastured grassland of western Europe, which is determined not only by climate and soil but also by specific and continuous operations of man. These formations are called *anthropogenic* (produced by man, from Greek ἄνθρωπος, man, and the root γένω, produce). The human activity involved—burning, pasturing, mowing, etc.—has to be taken as one of the constant factors of the habitat; indeed the human activity is here the differentiating or *master factor* (see p. 33).

A general feature of all formation-types is that the *dominant plants* of the formations belonging to each are characterized by particular *life forms*. By dominant plants we mean those plants which give the community its characteristic appearance or physiognomy and which also largely control its structure, for example the trees of a forest, the low-growing under-shrubs of a heath, or the typical grasses of a meadow. The term 'dominant' usually refers to the species which occurs with greatest frequency in the most important stratum of a plant community; in some instances, however, the species which exerts most influence on the other species is not the most frequent, and in some communities no one plant dominates. By life form is meant the type of plant body, with which is associated its life history. For example the deciduous broad-leaved tree (oak, elm, ash, beech, etc.) is a well-marked life form; the evergreen needle-leaved tree is another; a third is the perennial herb with a persistent underground stem and leafy aerial shoots which arise from fresh buds every spring and die down in the autumn; a fourth is the annual herb whose whole vegetative body dies every season, the species being continued from year to year by means of seed; and so on. The oldest and most obvious division of the life forms of the higher plants is into trees, shrubs and herbs, but each of these includes several distinct life forms—the herbs a great number. An account of life forms is given in Chapter 5.

The European summer forest formation, the native woodland over much the greater part of the British Isles, is a north-western extension of the Continental forest which is dominated by species of deciduous broad-leaved trees—a life form that is suited to the climatic conditions of western and central Europe. In central and northern Scotland this formation is partly replaced by pine forest, which may be regarded as a south-westward extension of the Scandinavian coniferous forest dominated by evergreen needle-leaved coniferous trees suited to the northern climate where snow lies for the long period of winter.

In most of the British area formerly occupied by deciduous forest this has been replaced by the semi-natural formation of pastured grassland dominated by the meadow-grass life form—besides of course the arable crops and plantations of alien trees, which are neither natural nor semi-natural. Within the same area also there are several edaphic formations, such as those of freshwater plants, reedswamp and the maritime formations of salt

marsh, sand dune and shingle beach, besides others, like the heath formation, which are determined partly by soil and partly by climate. Of these
last, the bog or 'moss' formation is another good example. In the very wet
climate of western Scotland and western Ireland, great tracts of flat or
gently sloping undrained or poorly drained ground are covered with *blanket
bog*, dominated sometimes by bog moss (*Sphagnum* spp.) and sometimes
by members of the sedge family. The blanket bog is clearly dependent on
the wet climate, but wherever adequate drainage frees the soil from the
waterlogged condition, quite different vegetation appears, so that adjacent
well-drained slopes are occupied by heath, scrub, or even woodland. The
bog or moss formation, dominated by the same or similar plants, also
appears in less wet climates under special conditions, e.g. where the soil is
waterlogged and also acidic, as in constantly wet hollows in heath (p. 27)
and in the so-called 'raised bogs' (see p. 65).

THE PLANT ASSOCIATION AND CONSOCIATION

The actual units of natural and semi-natural vegetation which we find in
the field belonging to one of the plant formations are of various nature and
status. The largest of them, which are dominated by particular species, are
called plant *associations* and *consociations*, the former dominated by more
than one species, the latter by single species. Thus an oakwood, dominated
by one of the oaks, *Quercus robur* or *Q. petraea*, or a beechwood, dominated by the common beech (*Fagus sylvatica*), is a consociation, while a
wood dominated by more than one kind of deciduous tree is an association.
Oakwood, beechwood and mixed wood all belong to the European deciduo-
ous summer forest formation characterized by the dominance of the
deciduous tree, but each by particular species of tree. Similarly the European reedswamp formation shows consociations dominated respectively
by the common reed (*Phragmites communis* (*australis*)), the reedmace or
bulrush[1] (*Typha latifolia* and *T. angustifolia*) and the bulrush or great reed
(*Schoenoplectus lacustris*). All three show the typical 'reed' life form.
Sometimes they grow intermingled (association), sometimes separately
(consociations).

All species found in an association or consociation other than the
dominants may be called *subordinate* species. These are strongly influenced
and often largely determined by the presence of the dominants, and sometimes because the growth and decay of the dominants produces a special
kind of humus or otherwise affects the nature of the soil. A well-defined
association or consociation contains subordinate species (1) strictly confined to it (*exclusive species*), (2) seldom found outside it, or (3) at least
found more often within than outside the particular community (lesser

[1] The name 'bulrush' is sometimes given to *Schoenoplectus* (*Scirpus*) *lacustris*, sometimes to *Typha*. The former use seems more 'correct' but the latter is more common.

degrees of exclusiveness). Species which are present in every example or nearly every example of an association or consociation, whether they occur outside it or not, are called *constants*. Both exclusives and constants are known as *characteristic species* of the community. Species which occur as often outside as within the community, and those which are found quite rarely and casually within its limits, are known respectively as *indifferent* and *casual* species.

The constants of an association or consociation can be determined by comparing the complete lists of species from a sufficient number of examples, the exclusive species only by comparing the lists with similar lists from other communities. The complete list of species, made by combining the lists from a great number of different examples, gives the *floristic composition* of the association or consociation for the area from which the examples were taken. Thus the association or consociation is characterized first of all by the life form of the dominants, i.e. by the formation-type to which it belongs, and then, floristically, by the dominants, the characteristic species, including constant and exclusive species, and the whole floristic composition.

The technical name of a consociation is formed by the stem of the generic name of the dominant with the suffix *-etum*, followed by the name of the species in the genitive case: thus the beech consociation is Fagetum Fagi sylvaticae, shortened to Fagetum sylvaticae. Where there can be no possibility of mistake about the species the consociation may be called simply Fagetum. Associations are designated by linking the names of the dominants, e.g. beech–oak or *Fagus–Quercus* association.

HABITAT AND ECOLOGICAL FACTORS[1]

The *habitat* of a plant community may be defined ecologically as *the sum total of the conditions of environment (ecological factors) which are effective in determining the existence of the community in that place.* Thus the general habitat of one of the great climatic plant formations is coextensive with the area covered by the corresponding climate which stamps the vegetation with the characteristic life form of the formation, closely suited to the prevalent climatic complex.

The European summer forest formation may be taken as an example. The characteristic life form of this summer forest is the deciduous tree, which drops its leaves in autumn and produces a fresh crop in spring. The warm and not too dry summer is favourable to the functioning of the rather delicate leaves of the deciduous summer-forest trees. In winter the soil is often too cold for the active absorption of water by the roots of the tree, and if the foliage persisted through the winter the tree would run the

[1] A detailed account of habitat is given in Part IV (Chapters 14–18).

risk of being killed by the ill-protected leaves continuously transpiring water which could not be replenished from the soil. The trunk, branches and twigs are covered with a bark and the winter buds with waterproof bud-scales which prevent any considerable loss of water from the tree during the winter. Thus within this particular climatic region deciduous summer forest is the natural type of vegetation in the absence of other factors definitely adverse to the growth of deciduous trees.

Such factors may be marked local variations of climate within the general climatic region, for example in places exposed to particularly violent winds where forest trees may be reduced to the condition of wind-cut scrub or prevented from establishing themselves altogether. Other adverse factors are of a different kind, and of these the most important is the nature of the soil (*edaphic factors*). The climatic plant formation of deciduous trees, in one or other of its associations or consociations, can establish itself on a wide range of soils, but there are some on which it is replaced by other formations suited to the particular soil conditions. Thus in western Europe deciduous forest does not easily colonize very shallow soils over rock and also the drier and more sterile sandy soils, where it is replaced by heath. And not only are trees unable to colonize the submerged soils of bodies of water, but also the characteristic climatic dominants such as beech and oak cannot establish themselves as dominants unless the soil is free from continuous waterlogging. In the shallow water on the edges of lakes there is, instead, reedswamp, and, on the soils saturated though not covered with water, marsh, fen or bog vegetation, or particular trees or shrubs which can tolerate these conditions such as alder and willow. Deep water, reedswamp, fen and bog, each has its own characteristic life forms, well suited to the particular conditions in which the plants grow and widely different from the life form of the deciduous forest tree, though all exist in the same general climate.

Besides climatic and edaphic factors there is a third class, *biotic factors*, due to the effects of other organisms, and those arising from human activity (*anthropogenic factors*), which bring about much of the semi-natural vegetation discussed in Chapter 2, very largely as the result of the continuous grazing of man's flocks and herds. The importance of man's influence on vegetation may be illustrated by reference to the Scottish highlands where, as a result of extensive animal grazing and repeated burning of forest and moorland, regeneration of the native pine forest is severely threatened (McVean 1963).

Particular types of climate like those determining the great climatic formations are, as a rule, far more widely extended over great continuous areas of the earth's surface than the variations of soil and soil water which determine formations like reedswamp, or the conditions of constant pasturing or burning which determine formations like the English chalk pasture or many of our heaths. Thus within the areas of the great climatic types various edaphic and biotic factors give rise to edaphic and biotic formations dominated by life forms very different from those of the climatic formations. Such local formations are by no means always limited to the area of one climatic formation. For example the reedswamp formation type, with its tall reedlike stems or leaves and rhizomes creeping in the soft mud or sometimes floating on the surface of the water, occurs in tropical and subtropical as well as in temperate climates, because the edaphic factor of waterlogged soil overrides the climatic factors. Reedswamp is, however, represented by different associations and consociations, dominated by different species in various parts of the world.

Thus we can recognize, in the case of any formation, determining or *master factors*, whether climatic, edaphic or biotic, which separate one formation from a neighbouring one. For instance a reedswamp on the edge of a river may be adjoined by a strip of pastureland; above this, on a gently sloping hillside, is oak forest. The master factor of the habitat separating the pasture from the wood is biotic, namely the grazing, which prevents the seedlings arising from acorns falling in the pasture from growing into trees. The master factor which differentiates the reedswamp from pasture and forest alike is the high water level (normally above the soil surface) at the edge of the river. The general features of the climate (rainfall, humidity of the air, temperature, etc.) permit the existence of all three formations, but are primarily expressed by the forest which is not exposed to the other (limiting) factors while the other two are differentiated, the one by biotic, the other by edaphic factors.

The existence of associations and of separate consociations within a formation often depends upon minor differences of habitat. Some parts of the general habitat may be rather more suitable for one dominant and some parts for another, while other areas again are intermediate, showing no great advantage for any one of the dominants, so that mingling of the dominants (*co-dominance*) occurs. Sometimes one of the dominants completely occupies an area to the exclusion of the others, though the habitat seems equally suitable for all. This may be due simply to the fact that the occupying dominant got there first, and prevented others from invading. Such relations may exist in some areas between oak and beech in the summer deciduous forest, and between bulrush, reedmace and common reed in reedswamp. There are many forests in which beech and oak flourish side by side, though beech is excluded from some habitats (heavy wet soils) in Britain which oak can occupy, and oak from some shallow soils which beech can colonize. Bulrush and reedmace tend to occupy deeper water than the common reed, but the three dominants are often mixed.

The subordinate species of an association often occur throughout segregated consociations as well, but there are frequently some species peculiar to each consociation because the local characters of the habitat (e.g. of the soil) which suit the dominant also favour particular subordinate species, or because of the direct effect of the dominant on the subordinate vegetation. Thus the shrubby and herbaceous vegetation of oakwood and beechwood contain most of the same species, but the deeper shade of beechwood together with the shallow-rooting habit of the tree tends to exclude shrubs from the undergrowth and also results in a more limited ground flora than that characteristic of oakwoods; while the occurrence in beechwoods of other species which are not found or are much less commonly found in oakwoods is associated with the more persistent humus and with the particular soils commonly occupied by beechwoods in this country (shallow chalk soils and sandy soils).

STRATA, OR LAYERS OF VEGETATION

The great majority of associations and consociations, all those in fact whose dominants are tall plants, have distinct *strata* or *layers* of vegetation below the dominant stratum. Thus an English wood commonly has four strata: (1) the tree stratum, (2) the shrub stratum, (3) the field or herb stratum, (4) the ground or moss stratum. Some woods have two field strata—consisting of tall and short herbs respectively, and there is sometimes a second stratum of lower trees beneath the shade of the dominants. Stratification is more complex in lowland tropical rain forest, where often three, and less commonly four or more, tree strata may be recognized (Fig. 3.1), although they may not be sharply delimited. The tallest trees are frequently scattered and rise above the general canopy. In the dense shade, the forest floor may be largely bare, or herb or dwarf palm layers may be present.

Each stratum has an environment, or habitat, which differs from that of the others. Thus the crowns of the trees of an English wood are exposed to full sunlight, and often to considerable wind, while all the other strata are more or less protected from both. The protection increases as the soil level is approached, so that the shoots of the lower strata are not only increasingly shaded, but are surrounded by a damper atmosphere, and enjoy a more equable temperature. The roots of the different strata also often have very different environments. The tree roots may occupy partly the soil and partly the subsoil or the cracks in the surface layers of underlying rock where this is near the surface; the roots of the herbs may be partly in the soil and partly in the surface humus; while those of certain herbs and the rhizoids of the mosses are confined to the surface humus.

An herbaceous marsh or fen community commonly has at least three strata—the uppermost consisting of the tall dominant grasses or sedges, at least one intermediate stratum of herbs, and a lowermost stratum of mosses and liverworts. The roots of plants of the different strata (or of different species whose shoots are in the same stratum) may occupy different layers of soil with, for instance, very different air, water and nutrient contents. A

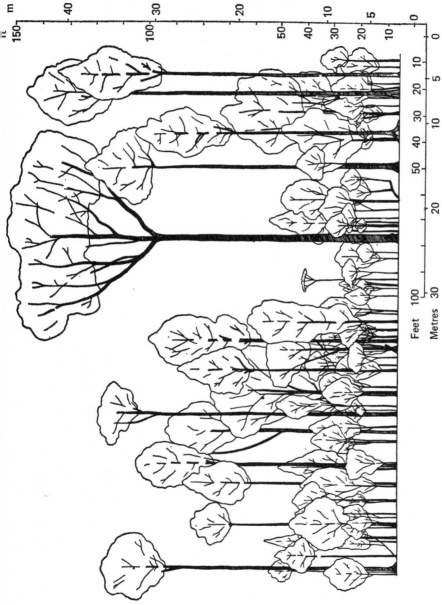

Fig. 3.1. A profile diagram of primary Mixed Rain Forest, Shasha Forest Reserve, south-western Nigeria. All trees 15 ft (4·6 m) high and over are shown. Three strata may be recognized : a dense storey of trees 25–50 ft (7·6–15 m), with closely packed crowns; an irregular middle storey of trees with small crowns; and a top storey of trees 120–150 ft (37–46 m), some (one shown) with umbrella-shaped crowns. The lowest storey grades into the shrub stratum (not shown). The belt represented is 25 ft (7·6 m) wide. (From P. W. Richards, *J. Ecol.* 1939, courtesy of British Ecological Society.)

grassland community is similarly stratified. In a comparatively damp climate like that of Great Britain the lowermost stratum of many communities consists of mosses.

The different strata of an association are in a certain sense distinct communities (see also p. 39). Each has a floristic composition, dominants, and a 'structure' of its own; and the species of each often belong to quite distinct life forms. Each stratum, as we have seen, has a habitat which differs, sometimes very widely. from those of the others. Consequently the different strata must always be considered separately in ecological study. On the other hand, the structure and often the existence of the lower strata depend upon the existence and nature of the upper ones. The shrubs of a wood will be much less dense if the tree canopy is heavy. The field stratum may scarcely exist if the shrub layer is very dense. Thus the different strata of a plant association are to some extent comparable with 'social strata', classes, or castes of a human community, for we find the same differences of habitat within the common habitat of the whole community, and a similar dependence between one and another. Each requires separate study from the sociologist, but the whole community to which they belong forms the essential primary unit in human populations and in vegetation alike.

THE SEASONAL ASPECTS OF A COMMUNITY

In a climate with well-marked yearly seasons different species of a community come to the height of their vegetative growth, flower and fruit, at different periods of the growing season. These activities of different species are scattered throughout the whole season, but the species tend to fall into distinct seasonal groups. In the British deciduous woodland, for instance, there are five such seasonal groups of species, and the flourishing of each gives a distinct seasonal *aspect* to the community. Thus we can distinguish (1) the *prevernal aspect* of early spring (usually March and the first part of April in southern England, but dependent on the weather of a particular season), marked by the coming into prominence of such plants as dog's mercury (*Mercurialis perennis*) (Fig. 3.2a, b), lesser celandine (*Ranunculus ficaria*), wood anemone (*Anemone nemorosa*), and primrose (*Primula vulgaris*); in (2) the *vernal aspect* (end of April and May), the trees come into leaf and flower, substantially reducing the amount of light reaching the woodland floor, and in the ground vegetation the bluebell (*Endymion non-scriptus*), stitchwort (*Stellaria holostea*), yellow archangel (*Galeobdolon luteum*) (*Lamiastrum galeobdolon*) and others flower; in (3) the *aestival* or *summer aspect* (June–August), a number of different species become prominent, superseding those of the vernal aspect, some of which disappear; (4) the *autumnal aspect* (September–November) shows very few or no fresh flowering plants in the woodland, but is marked by the appearance of many fungi and increased prominence of bryophytes; (5) the

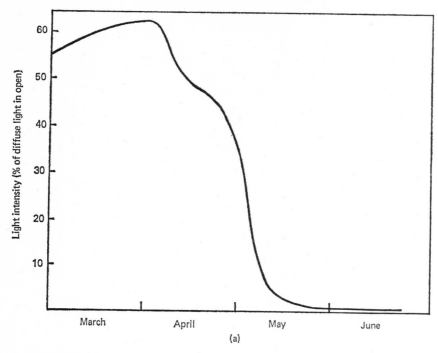

(a)

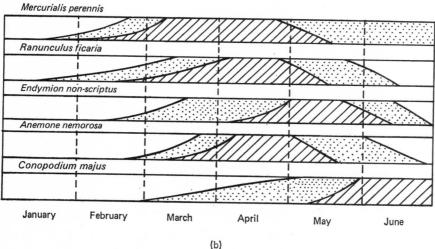

(b)

Fig. 3.2. (a) Seasonal changes in light intensity in an oak–hornbeam wood. (b) The period of foliage and flower production of common species of the shade flora. The onset of leaf and flower formation is shown, and the development to full leaf and bloom, followed by senescence. The vegetative condition is shown by stippling and flowering by hatching. (After E. J. Salisbury, *J. Ecol.* 1916, courtesy of British Ecological Society.)

winter or *hiemal aspect* (December–February) is marked by the flowering of winter aconite (*Eranthis hyemalis*) and snowdrop (*Galanthus nivalis*), neither British native plants, but often naturalized in woodland close to human habitations. The winter is of course relatively a resting period for most of the vegetation, but the underground parts of many plants continue to grow slowly.

We must be careful to distinguish the growth and activity of the leafy shoots of a plant from its flowering, though these are sometimes contemporaneous. The hazel (*Corylus avellana*) flowers in the hiemal and prevernal aspects, but its leaves do not appear until the vernal. Dog's mercury (*Mercurialis perennis*) flowers in the prevernal or early part of the vernal aspect, while its leaves are active in the prevernal and vernal and right through the summer aspect (Fig. 3.2b). The meadow saffron (*Colchicum autumnale*), on the other hand, produces its leaves in the prevernal aspect, but its flowers do not appear until the autumnal, long after the leaves have disappeared. Under the different aspects we have, therefore, to note what plants are in a state of vegetative activity, as well as what plants are in flower.

Deciduous woodland is characterized, in its field stratum, by rich prevernal and vernal vegetation, though it also contains many aestival species; grassland is mainly vernal and aestival, but is active to some extent throughout the year according to the weather; while heath, fen and salt marsh bear essentially summer vegetation continuing into the autumn.

THE PLANT SOCIETY

Within an association or a consociation certain species other than the general dominants form communities of lower rank. These *local* communities are called *societies*. Examples are societies of wych elm in an ashwood, of ash or alder in the damper parts of an oakwood, of horseshoe vetch (*Hippocrepis comosa*) in the chalk grasslands of southern England, and of mosses, such as species of *Polytrichum* and other genera, on heaths.

A society has usually a single dominant, which may occupy the ground to the exclusion of the association dominants. Sometimes the subordinate species within the limits of the society differ markedly in relative frequency from those of other parts of the association. Thus a society of a tree, such as the sycamore, casting heavy shade in an oakwood, will reduce or cut out many of the species common under the light oak canopy. Sometimes subordinate species occur in a society which are not present or are quite rare in other parts of the association: for instance the moschatel (*Adoxa moschatellina*) in the societies of dog's mercury (*Mercurialis perennis*) in the Derbyshire ashwoods. This may be because the society dominant occurs only in those parts of the association where the habitat is locally different (*habitat societies*), or again because the growth of the society dominant itself creates new local conditions which lead to the great abundance of certain subordinate species, or to the appearance of others not found elsewhere in the association.

The society dominant is a subordinate species when we consider the association or consociation as a whole, but within the society the other species may be subordinate to it. A society, as has been said, is 'a dominance within a dominance'. When highly organized, it gives a repetition of a structure of a consociation in miniature. In some instances, such societies represent fragments of a different formation, as patches of heath on sandy banks in a wood where the illumination is sufficient. The most slightly organized (lowest) form of society is represented by a mere local dominance of a subordinate species of the association (sometimes due to a chance aggregation of seedlings in one place, often to social vegetative growth) without change in the average distribution of the other subordinate species, except in so far as the social species excludes others by mass growth. The most highly organized type of society (approaching the character of a consociation) is represented by a characteristic list of subordinate species differing from that of the association at large, e.g. a society of cross-leaved heath (*Erica tetralix*) on the damper parts of heaths; of rush (*Juncus*) on the wetter parts of grasslands on heavy soils (here, as with a patch of heath in a wood, the society dominant has the life form of a different formation, corresponding with the marked local difference of habitat).

Stratum Societies

Societies of a stratified consociation may involve all the strata, i.e. within the society every stratum may show different frequencies of the species of the stratum, and perhaps some different species from the rest of the consociation. This occurs, for instance, in a wood where the dominant tree of a society has a marked effect on the habitat by casting greater or less shade, or by producing a different kind of humus from the consociation dominant. But other societies may be confined to one or two strata, e.g. the herb stratum, or the herb and moss stratum of a wood; or, again, the shrub stratum alone, e.g. societies of hawthorn. Societies not involving all the strata may be called *stratum societies*. The herb societies of woodland are often numerous and varied, and may or may not be correlated with the habitat conditions. It has been shown in the case of a Cambridgeshire wood (Adamson 1912) that the societies of dog's mercury, wild strawberry, and meadow-sweet are closely connected with the summer water content of the soil, and to a less extent with light. In Hertfordshire woods, societies of bracken and wood anemone, of lesser celandine, dog's mercury, etc., have been shown to depend on soil moisture, humus content and soil acidity (Salisbury 1916, 1918).

Groups of plants of similar life form often exist together in fairly uniform ecological conditions: for example, the cover of epiphytic lichens and bryophytes on the bark of trees. These distinct groups of plants of a complex community form what are called *synusiae*, assemblages which are often those of stratum societies, but characterized, not on a stratum basis, but on the basis of life form. An advantage of the latter basis is that it is possible to distinguish, for example, a synusia of woody lianes from a synusia of supporting trees.

'Aspect Societies'

Societies confined to one seasonal aspect of an association are sometimes called 'aspect societies'. But when a given patch of ground is occupied by the shoots and leaves of one or more species of one aspect (e.g. the pre-vernal), and by quite other species in another (e.g. the aestival), we must be careful how we refer to these as two different 'aspect societies', for we must remember that the underground parts of the species not in evidence on the surface during one aspect are nevertheless present all the time, and may influence the development of the species that happen to be conspicuous at the moment. A society composed of species whose aerial parts vegetate actively in different seasons or 'aspects' has been called a 'seasonally complementary society'. Each case must be decided on its own merits. The criterion of a distinct community is individuality.

THE CLASSIFICATION OF COMMUNITIES

As we have seen, the units of vegetation outlined in this chapter vary in size from the very extensive plant formations to the small, local communities referred to as societies. The essential feature of these collections of plants, of whatever size, is that they have a certain unity or individuality, and it is this which makes the community a recognizable entity. Many grades can, however, be found of the distinctness of communities, some communities being very readily recognizable as characteristic groupings of plants, while others are less well defined.

The limits of communities, and in particular the boundaries between adjoining communities, may similarly be sharp and clear-cut, or they may be diffuse. If, for example, two very different soil types adjoin one another, the boundary between the contrasting communities developed on these soils may be perfectly definite, and can often be easily recognized on the ground from the distinct change in vegetation. Sharp limits between communities may also arise as a result of fire or cultivation. More usually, however, changes in soil features and other factors of the environment which bring about the development of different communities are gradual and there is an appreciable transition zone or *ecotone* between two adjoining vegetation types. Fluctuations of the ecotone may sometimes be recognized, as, for example, between the marsh vegetation at the edge of a stretch of water and the adjoining terrestrial vegetation. After a succession of dry years, some species typical of the dry land vegetation may become established in the higher parts of the marsh; conversely after a succession of wet years, marsh plants may extend to terrain previously occupied by the dry land vegetation.

In some types of vegetation ecotones may be very wide and the environmental gradients only slight. In such situations boundaries are difficult to recognize and the limits chosen in vegetation mapping are of an arbitrary nature. In such instances the vegetation may form a *continuum*, one type of vegetation passing almost imperceptibly into another and no two types of vegetation being repeated exactly. Good examples of such intergrading vegetation may be seen in tropical regions and especially in the rain forests. However, the classification of communities into an orderly arrangement of vegetation units—especially where this can be readily accomplished without much difficulty, as for most of the vegetation of Great Britain—is of great convenience and a useful objective.

The system of classification of vegetation described in this chapter is one which has been widely and successfully used in Great Britain for many years but there are other systems based on somewhat different criteria and some of these classifications have been extensively adopted elsewhere. In particular the system developed by Braun-Blanquet (1932) and others on the continent has gained strong support, and has been applied to some British vegetation (see Chapter 7).

Chapter 4

The Succession of Vegetation

Life never stands still: it is everywhere in a continuous process of flux and change. In the last chapter the units of vegetation were treated essentially as if they were static units with a definite fixed composition, structure and habit, but in reality they are constantly changing. We must in the first instance define the natural units which are actually formed in the course of the changes of vegetation in order to have something to work on. Broadly, we may regard these units—our associations, consociations and societies—as representing positions of relative equilibrium into which plants group themselves. Some of these units are more stable, others less stable: some, that is to say, remain essentially the same for a long time, for centuries or perhaps for many thousands of years, whereas others are very transient, giving place in the course of a few seasons to other communities, of different composition. Sometimes change is fluctuating, different species being here today, gone tomorrow, and back again the next day, so to speak. This is especially seen in the annual and biennial vegetation of open soil—sandy coasts, roadsides, waste places such as derelict building sites, neglected garden beds, and all ground that has recently been disturbed.

But besides such fluctuating changes there can generally be observed, through a series of years, a definite trend in one direction towards a position of equilibrium. All such progressive change is called *succession*.

We must at the outset distinguish two classes of change which bring about the succession of vegetation. If an area of ground is bare of vegetation, or if new bare ground is formed—as by emergence of land from the sea, by the drying up of a lake, by deposit of alluvium by a river, by the retreat of a glacier, by a deposit of wind-blown sand, or by the fall of talus from a cliff—plants will, in nearly all cases, begin to occupy it sooner or later. The first species to occupy the area will in most cases, however, give way to others, and these again to others until a state of relatively stable equilibrium is reached between the vegetation and its habitat. This kind of succession is called the development of vegetation (*autogenic succession*), and any particular example is called a *sere*, a term relating to all the developmental stages of the vegetation from the initial colonization of a bare habitat to the final stage of equilibrium with the environment.

But change may also be brought about by continuously-acting outside factors which are constantly altering the habitat, thus making it less suitable for the first occupants of the ground and more suitable for others, as

42

for instance a gradual change of climate, the gradual silting up of a lake, the increasing concentration of salt in the soil as an inland sea dries up in an arid climate, or the 'leaching' (gradual washing out) of soluble salts from the surface layers of a soil by the percolation of rain-water. All these changes, which are independent of the plants, gradually alter the habitat, and thus bring about changes in the vegetation (*allogenic succession*).

The habitat also changes as a result of autogenic succession, by the reaction of the plants upon it. As the individuals die the products of their decay accumulate as *humus* in the surface layers of soil, altering its physical structure and chemical nature, and increasing its water-holding capacity. Thus the habitat is usually rendered more favourable to plant life, and species which make higher demands on the soil than did the original pioneers are able to obtain a footing. In general, bigger plants progressively replace smaller ones; and in all the more favourable climates of the world, on the more favourable soils, woody plants come to occupy the ground and forest vegetation is ultimately established. As the vegetation becomes closer and taller the shelter increases and provides a habitat for subordinate species which could not enter the earlier communities, so that the eventual forest communities become highly complex.

Of course, these two kinds of change are not necessarily separated in natural successions. Developments are often modified by concurrent changes in external factors. For instance, the development of aquatic reedswamp and fen vegetation on the edge of a lake may be modified by the deposition of silt brought down by streams (Pearsall 1917, 1918). This can be shown to alter the course of development, i.e. to lead to the formation of communities differing from those which would have appeared if there had been no silting. But it is important to keep clear in the mind the distinction between the two classes of factors (autogenic and allogenic) which influence succession. An instance in which the latter alone is at work is the effect of a gradual continuous change of climate on a plant community. Thus a forest which represents the culminating stage of development from a certain climate at first existing, and is thus in relatively stable equilibrium with its habitat, may alter in composition, or disappear altogether, to be replaced by some other type of community, as the climate becomes, for example, progressively drier or colder. Many instances of long-term changes of this kind, brought about by changes in temperature or rainfall, are now known, for example, from studies of the plant remains preserved in peat formed since the last glaciation.

The early course of development of vegetation on bare ground differs entirely according to the nature of the initial habitat, in the first place whether it is submerged or exposed to the air, and if exposed whether it is wet or dry; for the plants which can colonize such habitats differ very widely. On submerged ground, as in the shallow water on the edge of a lake or pond, we have the series: submerged aquatic plants, aquatic plants with floating leaves, reedswamp plants. As the soil level is built up by accumulation of plant remains, by silting, or by both together until it reaches the surface of the water, fen or marsh will succeed reedswamp. Shrubs and trees which can tolerate waterlogged soil around their roots (in this country such trees as alder, willows and birch) often follow, and as the soil level is built higher above the water-level by accumulation of humus, so that the soil becomes drier and better aerated, other trees come in, and the series of communities may be completed by the climatic forest formation. Such a sere, originating in water, is called a *hydrosere*.

On a dry, bare habitat, such as an exposed rock surface or the talus from cliffs, the early stages are totally different. Lichens and terrestrial algae, together with rock mosses—plants which can themselves hold rain-water— are the pioneers on bare rock surfaces. These begin to disintegrate the surface of the rock, and with the decayed parts of their own bodies form a thin soil. This is colonized by other mosses which form thicker cushions, and the soil gradually increases enough to support herbs whose roots can function successfully in a thin soil. Then, as the soil grows thicker, more strongly-rooting herbs appear, and after a time the seedlings of shrubs and trees take a hold. Eventually, with the thickening layer of soil and humus, comes the climatic forest formation. A sere taking origin on a dry habitat is a *xerosere* and on a rock surface a *lithosere*.

On the talus of cliffs the interstices between the rock fragments form an initial habitat very different from the surfaces of the fragments; the seeds of herbs and even of shrubs and trees often germinate in these interstices and these plants successfully establish themselves at an early stage long before they can colonize the thin soil formed by the lichens and mosses on the rock surfaces. This depends partly on the weathering of the rock—on how much mineral is formed by disintegration and washed down into the interstices, and partly on how much humus is carried down by rain from the surface vegetation. On a talus formed of small fragments—gravel, down to the size of coarse sand—the succession is much quicker and herbs play the leading part as pioneers, binding the loose gravel, forming mats on its surface, and comparatively quickly producing a suitable soil for the establishment of shrub and tree seedlings. On damp sand or silt, such as is laid down by a river in flood, colonization and succession are still quicker owing to the relatively favourable soil that is presented to the plants from the outset: herbs or even shrubs and trees begin to colonize the area at once.

The early stages of the development of vegetation thus differ quite radically according to the nature of the initial habitat, and this influences the course of succession for a considerable time. The *tendency* is always ultimately to develop the most complex community with the largest dominants that the climate permits, whatever that may be, and for this reason such a community is called the *climax community*, because it represents the final pitch of development which the particular climate allows. The complete sere from bare ground to the climax community is called a primary sere or *prisere*.

The climatic climax, or climatic formation, however, as we saw in the last chapter, is not developed on all areas within the corresponding climatic region. The nature of the soil, the presence of permanent bodies of water above or nearly reaching the ground-level, as well as other local but constant physical factors of various kinds, may altogether prevent it, and so may the existence of a permanent human factor like pasturing or periodic burning. Not only so, but various distinct communities, for example different forest associations or consociations usually with the same general life form but dominated by distinct species of tree, may develop on different types of soil or under different physiographic conditions, such as marked differences of exposure, within the same climatic area. Each of these forest communities may be in apparent equilibrium with its own particular environment and none may show any tendency to give way to another, so that we must recognize several different climax associations or consociations representing the climatic climax formation (*polyclimax*). This phenomenon must of course be distinguished from the arrest or diversion of the succession so that a formation is established quite different from the climatic climax. The community produced by the arrest of the succession in a stage marked by the dominance of a life form clearly 'inferior' to that of the climatic climax (for example, grassland or heath instead of forest) has been called a *subclimax*. But in a great many instances, probably to some extent in all, the arrest also leads to *deflection* of the sere, producing a community which does not exactly correspond (though it does correspond in a general way) with any stage of the prisere. This is notably the case where the arresting factor is human intervention, e.g. the regular mowing of fen in the primary hydrosere, or the continuous pasturing of chalk grassland in the xerosere on chalk soil which precludes succession to woodland. The 'mowing fen' and the chalk pasture are quite distinct stable communities, not identical with any stage in the respective priseres of fen and chalk down. They may be called *plagioclimaxes* (Greek πλάγιος, slanting, sideways) and the short sere, which leads to a plagioclimax by deflection of the prisere, may be called a *plagiosere*.

We use the term *climax* for any plant community which is stable in the sense of being in equilibrium with all the existing conditions, of whatever nature, to which it is subjected. In addition to the climatic climax we thus recognize *edaphic* and *biotic climaxes*, and many at least of these are, as we have seen, plagioclimaxes. Usually biotic climaxes result from the activities of man and so are termed *anthropogenic*.

When a community is not in equilibrium but is a *phase* in development leading to a climax it is a *seral community*. Seral communities corresponding in rank to associations, consociations and societies are often called *associes*, *consocies* and *socies* respectively.

In a country like Great Britain, where man has modified the spontaneous vegetation so that most of it is what we have called 'semi-natural', we can rarely find those long series of stages of development from bare habitats to the climatic climax which we have outlined above, and which we can study in regions of the world approximating to the virgin condition. We find instead a patchwork of communities from the pioneer communities of bare areas to the climatic climaxes, nearly all modified in various ways by man or his animals, and mixed with areas of sown or planted crops. All of these, if left to themselves, would progress towards the climatic climax on the more favourable soils, or to some edaphic climax on special types of soil; but man is constantly stopping or modifying the development or throwing it back to some earlier stage. Where he has introduced a more or less permanent modifying factor or set of factors, we have biotic (anthropogenic) climaxes or some stage of development towards them.

All development of vegetation initiated, not on new ground, but by some modification or destruction of pre-existing vegetation, is known as *secondary succession*, and it is with secondary successions (*subseres*) that we have mainly to deal in a country like Great Britain. The course of a subsere is necessarily different from that of a prisere on new ground, because the starting-point is different and the time occupied to complete it is less. Instances are the clear felling without replanting of a wood, or the burning of a heath. The subseres most like priseres are those which are started by complete destruction of the original vegetation and its soil, as when stone is quarried or gravel dug, and the gravel pits or quarries are afterwards abandoned. In such cases the original soil is completely destroyed, and the colonization begins on bare rock or on a loose but purely mineral surface. A parallel case in the water succession (*hydrosere*) is the digging of an artificial pond.

It is obvious that the colonization of new ground or of bared or partially bared ground must depend on the species of plants available to colonize it. This involves three factors: firstly, the actual proximity of seed or spore parents; secondly, the means of migration, i.e. the methods by which the seeds or spores are carried (wind, birds, etc.); and thirdly, the suitability of the habitat for the successful germination and establishment of the young plants.

The first colonists of dry areas (terrestrial algae, lichens, mosses) are widely distributed species whose spores are carried considerable distances by the wind. In a sufficiently damp climate they arrive and establish themselves quickly, in a dry one much more slowly. The herbs which usually come next are frequently annual species, often weeds of waysides and arable land, whose original habitats, before the plants became 'weeds', were just such raw dry soils of the early part of a natural prisere. Such plants often have light seeds, or structures which aid dispersal by wind, such as hairs and plumes attached to the seed or fruit. They form transient communities which often shift from place to place. The later-arriving perennial herbs appear more gradually, frequently because their means of transport are not so good, so that they come, a few only at a time, at longer intervals. They also commonly germinate more slowly, and require for germination more enduring moisture, so that seeds which arrived too early and fell on the surface of raw soil without humus would not grow into plants.

Finally, many of the trees and shrubs require much more favourable conditions for successful germination and for establishment, especially a deeper soil, so that they have no chance in the conditions of the habitat during the early stages of development (except for instance in the interstices of talus). And here the actual proximity of the seed parents becomes a very important factor in a country like Great Britain, where man has long ago destroyed most of the forest to which its climate is suited. The great distance of adequate numbers of seed parents from many areas which could be colonized by trees if seed was available, combined with the relatively poor means of dispersal of the climax dominants, oak and beech, is sufficient to account for the rarity of natural colonization of many suitable areas by these trees. Ash and birch have winged fruits and colonize more quickly over the country generally, while pine, which has winged seeds, often colonizes very freely if pinewoods or plantations are not too distant.

The earlier communities of bare ground are *open*, the individuals scattered here and there with stretches of bare soil between: the habitat is not fully occupied. They consist of comparatively few species, those suited to the special conditions of life in such situations—the lack of humus, the exposure, and the frequent dryness of the surface layers of soil.

The later communities become more and more *closed* (except where the soil or climate is very unfavourable, as in a desert, in which case the climax too remains an open association, at least above if not below ground), the individuals more numerous and the number of species increases as the habitat becomes more favourable for a greater variety of plants. Certain species become dominant, at first locally; and finally the dominance of a few—often only one or two—is established. In some climax associations, however, for example many grasslands, several species, generally of the same life form, are *co-dominant*. The community becomes definitely layered, species which can exist only in good shelter appear in it, and the final structure comes into existence. When this is once established it is difficult for new species to enter the association.

So far we have been considering the general succession of plant communities that can be recognized in a sere, but in addition to these progressive changes small-scale cycles of change are known to occur in many types of vegetation. Close and repeated examination of vegetation often reveals a recurring sequence of change in which one species is followed by another or a series of others, which in turn may be replaced by the first species, so completing the cycle.

Such a process may be detected on the surface of a growing peat bog, with its mosaic of shallow pools of open water and of hummocks at various stages of development. The hummocks may be actively growing, bearing bog moss (*Sphagnum* spp.), and the taller, older ones topped by ling (*Calluna vulgaris*), cross-leaved heath (*Erica tetralix*), cottongrass (*Eriophorum vaginatum*) and other species, eventually including the lichen *Cladonia arbuscula* (*sylvatica*), which are relatively intolerant of flooding. By the last phase the hummock has ceased to grow. In the wet conditions around existing hummocks *Sphagnum* grows rapidly, its organic matter accumulating quickly, and so new hummocks arise which ultimately over-top the old ones which become flooded. The younger hummocks, as they become drier, are in turn colonized by ling and other plants, the rate of hummock growth diminishing and the cycle being repeated. An upgrade or hummock-building phase can thus be recognized and a downgrade or degenerate phase in which the hummock ceases to grow and is replaced by a pool. We see here a small-scale succession constantly repeated, yet the mosaic of hummock and pool on the bog surface remains. This *regeneration complex*, as it is called, thus represents a series of phases dynamically related to one another. Preserved in the underlying peat is direct evidence for the changes, for examination of the plant remains in the peat reveals the cyclic repetition of the vegetation of hummock and pool.

Cyclical changes of a comparable kind can be recognized in other types of vegetation (Watt 1947a). In the Chiltern beechwoods, for example, the patchiness of the vegetation is related to a sequence of phases (Watt 1925), in which bare ground beneath saplings gives place to a phase in which the wood sorrel (*Oxalis acetosella*) is an important component of the ground flora and still later a phase in which the bramble (*Rubus fruticosus*) is the major constituent. Finally, on death of the old trees, there is a gap phase in which regeneration of the beech may take place. The sequence is closely related to the changing age and size of the dominant trees and their effects on the habitat and associated vegetation. The upgrade phase terminates in the death of the trees, and the downgrade phase with their decay and the early gap phase, in which light-demanding plants enter the sequence. In some instances the cycle of change may involve a phase dominated by another species of tree; for example, birch may colonize the gaps in beech woods and by its growth substantially alter the surface soil. Later beech seedlings may establish themselves and eventually overtop the birch.

Further illustrations of the dynamic nature of plant communities are provided by other examples of regular and repeated sequences of change, as, for example, in grasses on acidic sands, and of the growth of bracken (*Pteridium aquilinum*) and of *Calluna vulgaris* (Watt 1947b; Anderson 1961); such cycles involving pioneer, building, mature and degenerate phases are probably of very widespread occurrence. Thus although the succession or directional change of vegetation to a climax community which has a strong individuality can be recognized, within this community, whose overall structure is preserved, there may be constant small-scale cyclical change.

Chapter 5

Life Forms of Plants

To complete our brief sketch of the nature, structure, and behaviour of vegetation and of the units that may be recognized within it, some further account of the part played by the life forms of the plants composing these units is necessary.

As we have seen (p. 29) the type of plant body, i.e. the life form, of the dominants is the essential character of a plant formation. In other words, a formation-type is characterized by the life form of its dominants—the great evergreen leathery-leaved trees of a tropical rain forest, the deciduous trees of temperate forests, the grasses of the prairies and steppes, the dwarf shrubs of subtropical deserts, and so on. The characters which have to be taken into account in classifying life forms are those which are of ecological importance in adjusting the plant to its habitat, and many of these characters reflect the morphology or *growth form* of the plant.

One of the most important of all is the nature of the foliage leaves, not only because they are the major photosynthetic organs of the green plant, but because it is mainly through the leaves that plants are continually losing water by transpiration to the air, and if they lose water more quickly than it can be supplied from the roots they will dry up and eventually die. The form and structure of leaves are therefore closely related to the prevailing humidity or dryness in which the plant lives. The drier the air, the more readily is the plant liable to lose water. Leaves may be large or small, thick or thin, of soft or leathery texture; they may have a thick waxy cuticle with stomata which close tightly under drought conditions and in this way restricting transpiration losses and conserving water, or they may be only slightly protected against rapid transpiration; they may be broad or narrow, or scale-like, or even absent altogether when their functions are taken over by the stems.

Another fundamental feature of life form is the number and nature of the shoots which the plant produces. For example, we have the forest tree with its bulky woody stem bearing numerous branches raised far above the ground; the shrubs, also woody but much less lofty, and often producing several ascending stems from ground level; and the herb, with its comparatively soft body generally of lower stature still and showing great variety of construction. Besides these three main categories of flowering plant body, the 'lower' plants such as mosses, liverworts, algae and fungi, show numerous life forms ranging from mosses and giant seaweeds of very different size but both with comparatively complex structures, to simple microscopic unicellular forms of algae and fungi. It is, however, with the life forms of flowering plants that we are here mainly concerned.

Herbaceous plants show the greatest variations of life form. Thus we have the simple erect annual plant with a single, or only slightly branched, stem, contrasting with the perennial plant which persists from year to year by means of a stem more or less deeply embedded in the soil. Sometimes this is a compact and erect 'rootstock' and sometimes a large creeping rhizome running more or less horizontally not far below the surface of the soil. The stems of some plants produce thin creeping axillary shoots on the surface of the soil (stolons or runners) or below the surface (soboles or slender rhizomes), and both give rise to fresh plants vegetatively by rooting and sending up aerial shoots at a short distance from the parent. Then there are the compact underground shoots called stem tubers, corms and bulbs, which serve the purpose of perennation—persistence from season to season —and often also of vegetative propagation by the production of buds, usually lateral and detachable, which grow into new tubers, corms or bulbs.

Finally we have the trees and shrubs which, by forming persistent woody aerial stems, carry their perennating buds high above the ground.

These various forms of herbaceous or woody plant body, together with the great variety of actual shoots and the numerous types of foliage leaf that have been already mentioned, give rise to a great number of possible combinations and variations, and several different classifications of life form based on different vegetative features of ecological significance have been proposed. The Danish botanist Raunkiaer suggested a classification based on a single principle which has general validity and has proved of great value. This is the position in relation to ground level of the vegetative perennating buds which (except in annual plants) give rise to the new shoots and thus carry on the life of the plant from year to year. In most climates there is an annual season unfavourable to plant growth—in hot countries the dry season, in temperate and cold climates the winter. During this season, the great danger is loss of water which cannot be adequately replenished from the soil. Trees and shrubs, whose perennial shoots are raised high into the air, are largely protected against loss by bark and perennating (winter) buds whose tightly enclosing scales are waterproof. Nearer the surface of the soil the danger is considerably less because the air is damper as the result of evaporation from the soil and because the strength of the wind is much diminished. The best protection is obtained, however, if the perennating buds lie on or in the surface layer of soil or are completely buried. Similar protection is afforded to marsh and water plants by the immersion of their perennating buds in water or in subaqueous soil.

Raunkiaer's classification of life forms (1934), of which the chief are shown in Fig. 5.1, distinguishes plants with perennating buds borne well above the ground level (broadly trees and shrubs) as *phanerophytes* (Greek φανερός, *phaneros*, visible, exposed). The taller trees, more than 8 metres (25 feet) in height, are *megaphanerophytes* and *mesophanerophytes* (MM); trees and shrubs between 2 and 8 m (6 and 25 feet) are *microphanerophytes* (M); and shrubs between 0·25 and 2 m (10 inches and 6 feet) *nanophanerophytes* (N) (Latin *nanus*, dwarf). Undershrubs or herbs whose perennating buds are raised into the air, but not more than 0·25 m (10 inches) above the surface of the soil, are *chamaephytes* (Ch) (Greek χάμαι, on the ground).

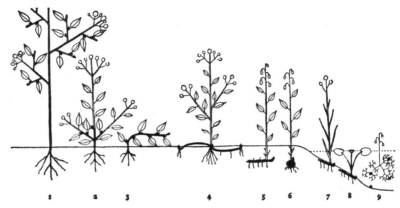

Fig. 5.1. Diagram of the chief types of life form based on Raunkiaer's classification: 1, phanerophytes; 2–3, chamaephytes; 4, hemicryptophytes; 5–6, cryptophytes; 7, helophytes; 8–9, hydrophytes.
    The parts of the plant which die in the unfavourable season are unshaded; the persistent axes and perennating buds are black. The sequence shown represents increased protection of the surviving buds, which are highest and most exposed in the phanerophytes. (From C. Raunkiaer, 1934, courtesy of the Clarendon Press, Oxford.)

Hemicryptophytes (H) (Greek ἥμι, half and κρυπτός, hidden) are plants whose buds are formed in the surface of the soil, and *geophytes* (G) or *cryptophytes* those whose perennating buds are buried more deeply and situated on a rootstock, rhizome, corm, bulb or tuber. Finally *therophytes* (Th) (Greek θέρος, *théros*, summer) are annual plants with a single growing season, but they include 'winter annuals' whose seeds germinate in the autumn, the seedlings surviving the winter in a vegetative form, often as a short stem with a rosette of leaves; the plants die after they have flowered and fruited in the next spring or summer. Water plants, whose perennating buds are situated under water, and marsh plants, with perennating buds in the soil or mud below the water level, are called *hydrophytes* and *helophytes* (HH) (Greek ἕλος, marsh) respectively.

The series MM, M, N, Ch, H, G, represents increasing protection of perennating buds as the result of closer approach to an eventual burying in the soil. Immersion in water or in subaqueous soil (HH) affords protection to buds comparable with burying in the earth (G), and therophytes perennate by means of seeds which are usually more resistant to desiccation than any perennating bud, so that these classes come at the end of the series of increasing protection.

Parasites, epiphytes (E) and stem succulents (S), put into special classes, cannot be brought into this series, but are of little or no significance in the British flora. Although a few parasitic flowering plants, e.g. mistletoe (*Viscum album*), dodder (*Cuscuta*) and broomrape (*Orobanche*), occur in Britain, there are no flowering plants which are true or obligate epiphytes (plants which live only on the surface of other plants but are not parasitic), such as exist in wet tropical climates. Several British species, however, which are normally rooted in the soil may grow successfully in the humus collected in the tops of pollard willows or in the crotches of trees, especially in the damp climates of the west.

All forest and scrub communities are dominated by phanerophytes; deserts are dominated by nanophanerophytes and chamaephytes (some desert regions by therophytes during the favourable season for vegetation), grasslands by hemicryptophytes, and much arctic vegetation by chamaephytes. The life form of the dominants, as we have seen, is the primary character of the great climatic plant formations.

Raunkiaer, however, was especially interested in another aspect of the distribution of life forms—the percentage distribution of the different classes in the *total flora* of a climatic region. It does not follow that the particular life form represented by the dominants of a climatic formation will also be the most frequent life form in that climatic region. Sometimes it is, as in the region of tropical rain forest where phanerophytes are not only dominant in the rain forest but also form the great majority of all the species of plants growing in the region. But this is not true of the deciduous summer forest region where phanerophytes are still dominant in the climatic forest formation, but are far outnumbered by hemicryptophytes in the total flora. Raunkiaer showed that what he called the '*biological spectrum*' of a region, i.e. the list of percentages of the different life-form classes represented in the total flora, was clearly related to the climate.

Most of Britain, like Denmark, belongs to the hemicryptophyte climate of north-western Europe (Table 5.1): only on the higher mountains do the chamaephytes increase in number until they approach the proportion found in the arctic regions, while the hemicryptophytes are still a very strong component. These life forms are successful here because the snow which covers the ground during the long winter serves as a protection to the perennating buds borne close above or on the surface of the soil. The buds burst into leaf and flower at the beginning of the short growing season

Table 5.1 The biological spectrum, showing the percentage representation of different life forms, for Denmark, Clova and Spitsbergen. (After C. Raunkiaer, 1934, courtesy of the Clarendon Press, Oxford.)

	Denmark	Clova, Scotland, under 300 m	Spitsbergen
Number of species	1084	304	110
Life Forms			
Mega- and mesophanerophytes	1	3	0
Microphanerophytes	3	2	0
Nanophanerophytes	3	4	1
Chamaephytes	3	7	22
Hemicryptophytes	50	59	60
Geophytes	11	7	13
Hydrophytes and helophytes	11	5	2
Therophytes	18	13	2

when the snow melts and the soil is rapidly warmed and well illuminated during the long hours of daylight.

The life-form classes are also useful in analysing the floristic composition of particular associations and consociations. Forests, the most complex of all plant communities, contain almost all the life forms of land plants (annuals and succulents are, however, rare in them), though the proportions vary according to the climate. Pasture consists almost exclusively of hemicryptophytes with a few geophytes, because buds above the soil level are eaten off. Heaths are characterized by the abundance of evergreen nanophanerophytes and chamaephytes, in relation to the mild winter climate. Permanent heaths are determined by some factor—climatic, edaphic or biotic—which prevents the establishment of trees. Communities inhabiting loose and soft soils of all kinds are marked by the abundance of geophytes with widely spreading rhizomes or soboles. Therophytes are abundant not only in many deserts but also in all open soil habitats, as in early phases of succession, on sea shores and sand dunes, in waste places and as weeds in all cultivated land.

Life form is, of course, primarily hereditary, though extremely severe conditions of life may force plants normally belonging to one life form into the class below by the killing of the upper buds: for example the wind-cut scrub of nanophanerophytic status formed on places exposed to violent winds may be of species which are micro- or mesophanerophytes in normal conditions. Life form represents the adjustment of the vegetative plant body and life history to the habitat factors, whether this is due entirely to heredity and selection or partly to the direct effect of the conditions of life.

Although classification on the position of the resting bud relative to the soil surface provides a series of groups of similar life form and related biology, not all ecologically distinct groups of plants can be distinguished by this criterion. Raunkiaer himself, as already noted, included a number of other life-form categories in his system, such as parasites and epiphytes, which may be abundant and conspicuous in tropical and subtropical vegetation. Moreover he used leaf size as a basis for classification, recognizing a series ranging from megaphylls, with leaf areas of more than 164,025 mm², through macrophylls, mesophylls, microphylls and nanophylls to the smallest category of leptophylls, with leaf areas of less than 25 mm². Leaf size is related to climate in that, in general, plants with large leaves are characteristic of warm, wet regions and those with small leaves of cold or very dry ones. However, soil factors too may be important in determining leaf size.

Other systems of biological classification may also be of ecological value, for example that which distinguishes, at one end of a series, trees with thick, unbranched or little branched stems bearing a terminal crown of large, compound leaves, such as in palms and tree ferns (*pachycaul*) and at the other the fine-twigged, much branched trees with small, entire leaves (*leptocaul*), such as our English elms. A tendency to a pachycaul habit is seen in the ash, with its pinnate leaves and thick stubby twigs. Another way of classifying plants is on the principal agent of seed dispersal: whether wind, animals or water for instance is most important. Plants suited to long-distance wind dispersal occur most frequently in the taller strata, and plants of pioneer communities usually have more efficient means of spread than plants of climax vegetation. As shown above, various methods of classification are possible, based on features of physiognomy (general appearance) or on functional characteristics. The categories of plants recognized are ecologically meaningful, although they are based, like those of Raunkiaer, on a single criterion, because these characters reflect important biological differences.

Chapter 6

Outline of British Vegetation

At this point we may usefully introduce a summary outline of existing British vegetation in order to provide an indication of the range which is represented. A full account is given in *The British Islands and Their Vegetation* (Tansley 1953), and well-illustrated descriptions of the chief types of British vegetation are also given in *Britain's Green Mantle* (Tansley 1968).

In Great Britain we have a varied assemblage of plant communities, most of them more or less modified, and many created by man. A certain number, however, notably many of the freshwater communities, those of the seaweeds, the maritime vegetation of sand dunes, shingle beaches and salt marshes, the vegetation of sea cliffs and of some of the remoter mountains, moorlands and bogs of the north and west, are but little, if at all, influenced by human agency; that is to say they would have been just the same, as far as we can tell, if man had never inhabited this island. But they are all, whether entirely natural, semi-natural or artificial, open to ecological study, though some will be more fruitful than others.

WOODLAND

Deciduous Summer Forest
The climatic climax of the plains, the valleys and the lower hill slopes of England, Wales and much of Scotland, as well as a large part of Ireland, is the deciduous summer forest formation of western and central Europe. In the British Isles it is dominated mainly by one or both of the two deciduous oaks, *Quercus robur* and *Q. petraea*; and on certain soils, especially on the chalk in the south-east and on the oolite of the Cotswolds somewhat further north and west, by the beech (*Fagus sylvatica*). The beech is now almost confined, as a wood-forming native tree, to the south of England, though it flourishes, sets seed, and may reproduce itself freely when planted in suitable situations in Scotland and Ireland. On the calcarous soils of the older limestones of the north and west, outside the area of the beech, the woods are dominated by ash (*Fraxinus excelsior*), though they are seldom very extensive. This ash consociation has not been described outside the British Isles, and even here most ashwoods are probably modern in origin. Indeed pollen of ash becomes abundant in the fossil record only after widespread forest disturbance by Neolithic man. Nevertheless numerous natural or semi-natural ashwoods occur on steep slopes on

Carboniferous limestone, often on scree or shallow soils with rock outcrops (Plate 1).

The other British trees which dominate widely distributed communities are the alder and the birches (see below). Of markedly gregarious trees the hornbeam (*Carpinus betulus*) plays a certain role in parts of the south-east, the area in which it is certainly indigenous, being even more restricted than the beech; and the yew (*Taxus baccata*), with a much wider distribution, is perhaps a climax dominant on some parts of the southern chalk. But both of these trees are better considered as society, rather than as consociation, dominants. The hornbeam tends to be subordinate to the oaks, the yew to the beech and sometimes to the oaks.

The native woods have very rarely been left untouched: only small fragments of virgin wood now exist in some of the remoter valleys, and often towards the altitudinal limit of the woodland. An example is Wistman's Wood on Dartmoor (Plate 2), where pedunculate oaks (*Quercus robur*) of great age, but highly dwarfed and with very thick trunks, are rooted among the granite boulders.

Great areas of woodland have long ago been cleared and converted to arable land or pasture. Indeed, the proportion of the country occupied by woodland, including plantations, is the lowest in Europe except for Ireland. In Great Britain only about 6·7 per cent of the land area is forested, compared with 28 per cent for the whole of Europe, and 29 per cent for the world. Of this forested land in Great Britain just over half is woodland of broad-leaved trees.

In the oakwoods which remain in the south, the shrub-layer, predominantly of hazel (*Corylus avellana*) with other shrubs associated, is almost always coppiced. A few standard oak trees only are present, not enough to form close canopy (Plate 3); or the coppice is left without standards. Both the oakwoods and the beechwoods of the south have certainly been largely planted, probably in many instances where the particular tree was formerly naturally dominant. Some beechwoods, however, and many oakwoods are almost certainly the direct descendants of natural woods. Natural regeneration of oak and beechwood at present occurs only here and there, so greatly have the natural conditions been altered.

The ash, which reproduces more abundantly and distributes its fruits more widely than either oak or beech, consequently springs from seed much more often, but where pure ashwood is found within the area of native beech it is generally to be regarded as a seral community or consocies.

The same is true of the birches (*Betula pendula* and *B. pubescens*) which, like the ash, are light-demanding trees unable to grow in full competition with oak or beech, and which produce abundant easily scattered tiny winged fruits; ready germination results in large crops of seedlings. In natural succession the ash commonly precedes the beech, and the birches the oaks, but this rule is not invariable. The actual relations in any given case depend largely on soil preferences—broadly the birches can range over a wide variety of acidic soils whereas the ash is more successful on basic soils—as well as on distribution factors, and are too complicated to discuss in detail here. The birches are very tolerant, not only of soil conditions but also of temperature, flourishing even in subarctic climates and reaching high altitudes (Plate 4).

In permanently wet places with neutral or basic soil the alder (*Alnus glutinosa*) appears to be the climax dominant, though it gives way to oak if the soil becomes progressively drier. Associated with it in the succession are the crack willow (*Salix fragilis*), as well as other species of shrubby willows, and often also ash and birch—the former only on neutral or basic soils. Certain alder woods, for instance in the region of the Norfolk Broads, are very probably essentially virgin; they contain trees of all ages, including saplings as well as old and decaying trunks.

As a whole, the British woods (apart from modern plantations of conifers, see p. 23) are semi-natural; that is they are the more or less modified descendants of original natural forest. This statement does not of course imply that there are not many plantations of native deciduous trees. The greater part of these are, however, on the sites of old woodlands, and since the indigenous trees have often been planted it becomes very difficult, or even impossible in the absence of accurate historical records, to say of any given wood that it is certainly planted or certainly the direct descendant of natural forest. A plantation of the natural dominant on its own soil will in course of time assume most of the characters of a natural wood (cf. p. 23).

Northern Coniferous Forest

In northern Scotland the oakwoods begin to thin out and eventually disappear altogether. In certain places in north and central Scotland there are native woods of the true Scots pine (*Pinus sylvestris* subsp. *scotica*). This region has a climate distinctly different from that of the rest of the British Isles, and is really part of the oceanic division of the region of north European coniferous forests, which are better represented in Norway. Associated with the pine of this region—and indeed far more widespread— are birches, the commonest forest trees of Scotland and the dominants of much climax woodland of the Scottish highlands. These pine and birch consociations of the north belong therefore to a different formation from the deciduous forests of the south, and their subordinate vegetation, often dominated by whortleberries (*Vaccinium myrtillus*, *V. vitis-idaea*), is much poorer in species. Many of these are the same as those of the southern woods, the hardier species of which, having penetrated farther north since the retreat of the ice, are able to stand the more severe climate; but the northern woods have also a few species of their own.

The common pine (*Pinus sylvestris*), which is a very widely distributed species in Europe and northern Asia, formed extensive forests in two post-glacial periods. In early post-glacial times the climate of southern England was subarctic: later on it became much warmer and forests, sometimes dominated mainly by pine, sometimes mainly by oak or other broad-leaved trees, spread over the country (Godwin 1956). The remains of these forests, largely in the form of pollen grains, are preserved in peat deposits. Possibly some of the existing southern pine may be descended from these post-glacial forests, but in the seventeenth and eighteenth centuries the common pine was extensively planted in England. On the light sandy soils here it has in many places spread from the plantations and established itself in pure subspontaneous woods or among the oak and birch, particularly in Hampshire, Surrey, Sussex and Kent; much of this subspontaneous pine was, however, cleared and used for pit props during the two World Wars. It is interesting to note that in Denmark and Belgium, as in England, there is now no native pine forest though the tree is extensively planted; but in these countries it disappeared only during the historical period.

HEATH FORMATION

The heath formation of north-western Europe is developed in wide stretches in western France and throughout the British Isles; it is also found in Jutland, Holland, southern Scandinavia and north-west Germany, mainly on sandy and dry acidic soil. The principal consociation is Callu-netum vulgaris, dominated by the common ling (*Calluna vulgaris*) which is often pure over considerable areas. The consociation Ericetum cinereae, dominated by the purple or bell heather, *Erica cinerea*, is confined to the western part of Europe. On the serpentine rocks of the Lizard in Cornwall, and in Brittany, Spain and Portugal, there is a consociation dominated by the handsome mauve-flowered species *Erica vagans*. On some areas the heath association is represented by a consociation of whortleberry or bilberry (*Vaccinium myrtillus*), a deciduous ericaceous under-shrub, which is also often mixed with *Calluna*.

The status of the heath community—its relation to the habitat and to other communities—is different in different places. On exposed high-lying peaty moorland where trees cannot grow, it is often the climatic climax. On sandy soils throughout the British Isles the heath community often occupies the ground, but here it will generally be succeeded by trees, especially birch and pine. Oak and beech sometimes follow, if seed is available. In other places it seems that forest cannot establish itself. This appears to be because of the failure of tree species to compete with the closed heath vegetation, in particular for nutrients. Successful establishment of trees can be achieved only after cultivation (e.g. deep ploughing) and removal of the heath species. Certain microbial interactions may also be involved, since extracts of *Calluna* peat have been shown (Handley 1963) to be particularly inhibitory to mycorrhizal fungi whose presence, in association with the roots of many kinds of trees, is essential for healthy and vigorous tree growth. Well-known mycorrhizal fungi of both coniferous and deciduous trees are almost totally inhibited by dilute extracts of *Calluna*.

Where features of the soil, which is often shallow and dry, prevent succession to forest, the heath community is an edaphic formation. It is certain, however, that many of our heaths are maintained in this condition not by soil character but by burning and grazing, and here the formation is biotically determined.

Callunetum requires an oceanic or suboceanic climate, and disappears as the Continental climate of central and eastern Europe is reached. It cannot grow on some soils, and does not survive heavy shading, so that it is dominant only in situations and on soils where thick continuous forest cannot become established. On the other hand it often maintains itself between the trees of an open birchwood which does not cast heavy shade. Grazing and trampling destroy it, and, subjected to these conditions, it gives way to certain types of grassland.

GRASSLANDS

The great bulk of the 'rough grazings' of Great Britain—excluding land which was formerly arable and has been sown with grass seed ('laid down' to grass)—are partly man-made associations, almost always due to pasturing though not to sowing. The chief exceptions to this origin of uncultivated grassland are certain maritime, submaritime, and mountain grass communities where edaphic or climatic factors, especially sea-salt and violent wind near the sea or at high altitudes, prevent the establishment of woody plants. For the rest it may be said that where grass grows, or rather where most of our meadow grasses grow, trees can grow; and that the cause of their absence is that their seedlings are eaten off and killed, while the dominant grasses of the sward, after being eaten down, constantly shoot again from buds on or in the surface of the soil. However, reduction of the rabbit population by myxomatosis has, of course, resulted in much less pronounced effects of grazing than those found when rabbits were plentiful.

This relation of forest and grassland can be clearly seen on many of the English grass 'commons', which still bear fragments of unfenced woodland. The part of such a common nearest a village, for instance, may be pure grassland. Sometimes it bears isolated bushes or patches of scrub consisting of spiny shrubs (gorse, hawthorn, blackthorn, bramble, briars) which the grazing animals avoid. The grass goes right to the edge of the wood and between the outlying trees which cannot regenerate, sometimes because any tree seedlings which became established would be eaten off, or because the compacted soil forms too firm and dry a surface for the seeds of the trees to germinate on successfully. Thus the area of woodland will constantly shrink and be replaced by pasture.

Within the shade of the wood the grasses of the open common can no longer grow. If the wood is small and the grazing on the common heavy, few or no woodland species will exist because the animals come right through it, eating off or trampling down any herbs which appear. But if it is larger and not much entered by grazing animals, woodland plants will be seen as soon as the marginal zone is passed; and young trees may be found where there is sufficient light and the surface soil is suitable for the germination of their seeds.

The actual communities dominated by grasses occurring on such pastured commons vary according to the soil. A widespread association of sandy soil, and of the somewhat similar soils of many northern and western hillsides whose rock is siliceous, is dominated by the common 'bent' (*Agrostis tenuis*) and the sheep's fescue (*Festuca ovina*), often with the sweet vernal grass (*Anthoxanthum odoratum*). This is a biotic (grazing) association developed where heath and possibly forest would develop if grazing were stopped. If the grazing is less heavy, the heaths (*Calluna* and *Erica*) are often locally or generally dominant. If the soil is distinctly acidic and peaty, the wavy hair grass (*Deschampsia flexuosa*) or the mat grass (*Nardus stricta*) becomes prominent or even dominant, and in damper places the purple moor grass (*Molinia caerulea*) is often dominant.

On heavier and 'better' soils with a higher content of soluble bases the common meadow grasses, such as *Poa pratensis, P. trivialis, Cynosurus cristatus, Lolium perenne* and *Dactylis glomerata*, form good pasture. The corresponding woodland is the pedunculate oak consociation (Quercetum roboris). Most of this 'meadow grassland' has, however, been laid down to grass, i.e. ploughed and sown, and forms what is called 'permanent grass', which is also a semi-natural community though of different status from the grassland created by grazing alone. Where there is much lime in the soil the oat-grasses (*Helictotrichon pratense, H. pubescens*) and the allied *Trisetum flavescens* often become prominent. On the dry soils of the chalk (and also on the older limestones) there used to be, typically, a closely grazed turf of *Festuca ovina* (sheep's fescue) and *F. rubra*, associated with a number of herbs characteristic of dry or highly calcareous soils, giving the well-known and highly characteristic association of chalk pasture.

The communities mentioned, or modifications of them, occupy between them the greatest extent of the semi-natural grasslands of Great Britain. Since they are all associations whose form and constitution are determined by grazing, they all give place to different communities if grazing is withdrawn.

In the last twenty years grazing pressures, particularly on marginal land, have been relaxed not only by myxomatosis but also because of changes in agricultural policy. As a result, there has been a tendency for a reduction in the number and diversity of species and the increasing dominance of taller-growing grasses.

The limestone grasslands of Derbyshire offer good examples of the effects of grazing and of aspect on the floristic composition. In Lathkilldale, where grazing has been almost non-existent since myxomatosis (1954), false oat-grass (*Arrhenatherum elatius*) has become increasingly prominent on the shallow soils overlying scree, particularly on the south-facing slope. In Cressbrookdale, where grazing by sheep and cattle has been maintained, there is a much greater diversity of species and the proportions of such grasses as *Festuca ovina* are also greater; *Arrhenatherum elatius* is absent from all grazed sites. In both dales, both north- and south-facing slopes contain the grasses *Helictotrichon pratense*, *Koeleria cristata*, and *Festuca ovina* and a few other species in common, but the rock rose (*Helianthemum chamaecistus*), thyme (*Thymus drucei*) and purging flax (*Linum catharticum*) are restricted to the dry, shallow sides of the south-facing slopes, while devil's-bit scabious (*Succisa pratensis*), and the grasses *Festuca rubra* and *Anthoxanthum odoratum* are restricted to the damper north-facing slopes.

Many areas of semi-natural chalk grassland in the south of England are reverting to scrub even more rapidly than limestone grassland further north, and in the south the grass *Brachypodium pinnatum* has increased substantially in the last twenty years. This tendency, together with increasing land-use for other purposes, is reducing the area of semi-natural grassland at an alarming rate (Plate 5).

Provided that seed of the appropriate trees is available, forest will ultimately develop, but the kinds of shrubs and trees that grow and the conditions under which they establish, as well as the corresponding changes in the soil, require detailed study.

The more porous soils show a general tendency in our climate to progressive 'leaching', i.e. washing out of soluble salts from the surface layers of the soil. This frequently leads to increasing acidity of the surface soil with a corresponding gradual change in the vegetation with lapse of time. How far and in exactly what way this external cause of succession affects the development of the vegetation to forest is a question on which we need more information.

FRESHWATER COMMUNITIES: MARSH, FEN AND BOG ('MOSS')

The next group of communities that we must notice are those developed in the succession of vegetation from freshwater to land, which takes place partly by silting and partly by the growth of peat at or near the water-level.

The freshwater communities themselves, consisting partly of algae, with mosses in some waters, and partly of flowering plants, present peculiar difficulties of classification. The most obvious rough division is into completely submerged plants, plants with floating leaves, and plants with predominantly aerial shoots. We know, however, that the richness or poverty of the water in dissolved oxygen, in various soluble mineral salts, and also in suspended material (silt), has a determining influence on these communities.

The types of vegetation borne by the alluvial and peat land produced on the edges of freshwater (lakes and rivers) are influenced by the same factors, and we may distinguish between marshland, fenland and bogland according as to whether the soil is formed mainly by silt, by peat containing considerable quantities of lime, or by peat very poor in lime and acidic in reaction. Marshland may approximate to fenland or to bogland according to the abundance or poverty of lime in the silt, and need not be dealt with here separately.

Fenland occupies the upper parts of the sites of old estuaries as well as the edges of certain lakes and the alluvial soil bordering streams. It is particularly well developed in East Anglia, where the rivers largely drain from the chalk and bring down 'hard' waters rich in lime. The greatest area of fenland lies between Cambridge and the Wash, but this is almost wholly drained and cultivated. The smaller area in East Norfolk (the 'Broads' region) is in a much more natural state.

The development of fen commonly starts from the reedswamp association, which may be regarded as the culmination of the series of aquatic communities proper. Reedswamp shows consociations of *Schoenoplectus* (*Scirpus*) *lacustris* (great reed), *Typha latifolia* (reedmace) and *Phragmites communis* (common reed). These plants may border deeper water in which floating and submerged aquatics occur (Plate 6). Reedswamp may also comprise smaller communities (consociations and societies) of other species, including the tall sedges (*Carex riparia*, *C. pseudocyperus*, etc.) and grasses such as *Glyceria maxima* and *Phalaris arundinacea*. As the submerged soil approaches the surface of the water by the continued accumulation of organic debris from the reedswamp plants, the landward edge of the reedswamp is gradually invaded by fen plants, i.e. plants whose shoots are subaerial instead of partly submerged. The two most characteristic fen dominants are *Cladium mariscus*, the stout saw-leaved fen sedge, and *Juncus subnodulosus*, the fen rush, though the common reed (*Phragmites communis*) and in some fens *Glyceria maxima* and *Phalaris arundinacea*, may remain dominant so long as the water-level remains sufficiently high. The two last-named are characteristic of reedswamps and fens in which there is abundance of mineral salts in the water. Such fens are also characterized by a greater variety of tall luxuriant dicotyledonous flowering plants in which the meadow-sweet (*Filipendula ulmaria*) and the umbellifer *Angelica sylvestris* may be conspicuous (Plate 7). The purple moor grass (*Molinia caerulea*) becomes dominant in some fens as they become drier and especially if *Cladium mariscus* is damaged by too frequent cutting.

The fen association, if not regularly cut, is soon colonized by shrubs and trees. Among the shrubs are *Salix cinerea, S. repens,* and other willows, buckthorns (*Rhamnus catharticus, Frangula alnus*), and the guelder rose (*Viburnum opulus*); and of trees, the birch (*Betula pubescens*), the ash (*Fraxinus excelsior*) and the alder (*Alnus glutinosa*) establish. Ultimately fen wood or 'carr' is formed, typically dominated by alder, but sometimes by birch or by alder buckthorn (*Frangula alnus*) as at Wicken Fen in Cambridgeshire (Plate 8).

Bog, as we have seen, differs from fen because it is developed wherever the water is very poor in bases, and it supports a wholly different plant community. The low levels of dissolved substances in the water mean a poor supply of essential mineral nutrients, and the bog plants characteristically exist under these *oligotrophic* conditions. In strong contrast are the fen plants, which grow under *eutrophic* conditions where supplies of nutrients are high, the water being rich in dissolved substances. Bog communities appear on the shores of lakes and pools in situations where the rock is deficient in lime and other bases so that the water is markedly 'soft'. This is well seen in Galway and Mayo in western Ireland where innumerable lakelets left in depressions after the retreat of the ice are often surrounded by vast areas of bogland. In a very wet climate, like that of western Ireland and western Scotland, this kind of bog covers great stretches of the low country, except where the local drainage is especially good, and is called *blanket bog*. It consists largely of bog moss (*Sphagnum*) and a number of other characteristic plants many of which, such as cotton-grass (*Eriophorum*), deer sedge (*Trichophorum caespitosum*), beak sedge (*Rhynchospora*) and black-headed sedge (*Schoenus nigricans*), belong to the sedge family (Cyperaceae). Some areas bearing blanket bog formerly supported the growth of pine, as shown by the exposure of sub-fossil stumps where peat has been cut and eroded (Plate 9). Blanket bogs also cover the elevated plateaux of hill masses such as Dartmoor, the Pennines, and many others in the north and west of Britain and in Ireland, where the climate is very wet. These plateaux are often covered by great sheets of glacial 'till' left by the retreating ice, and forming an impermeable substratum to the bog.

Bog may also be developed in fen basins where the climate is not too dry. Bog moss (*Sphagnum*) colonizes the fen vegetation above the level of the 'hard' (i.e. lime-containing) ground water, and forms cushions on the fen which are then colonized by other bog plants, giving rise to hummocks with wet hollows between them. An indication of the recurring sequence of vegetation change in the hummock–hollow complex has been already given in Chapter 4 (p. 48). The bog moss has a remarkable power of holding water and consequently its cushions can grow in height and extend laterally, depending entirely on rain-water. The lateral fusion of the cushions may result in the formation on the surface of the fen of a great lens-shaped bog which grows in height at the centre and whose edges extend so that the bog

1 Ashwood at Malham Cove, developed on the carboniferous limestone. Rock fragments form scree slopes which are partly stabilized by the trees.

M. C. F. Proctor

2 Wistman's Wood, Dartmoor. Oak woodland clothes part of the steep eastern side of the valley of the West Dart, while bushes of gorse (*Ulex gallii*) are on the slope in the foreground. In the distance is acidic grassland dominated by *Nardus stricta* and *Molinia caerulea* (Longford Tor is on the skyline).

May, M. C. F. Proctor

3 A coppice-with-standards woodland; the large trees of pedunculate oak (*Quercus robur*) are widely spaced, and there is a coppice shrub layer of hazel (*Corylus avellana*). Hayley Wood, Cambridgeshire.

February, M. C. F. Proctor

may come to fill up the whole fen basin. The substance of the lens consists of peat formed by the dead remains of the bog moss and accompanying bog plants which maintain themselves on the surface as the bog gradually increases in thickness and extent. This kind of bog is called *raised bog*, and its great stretches of reddish-brown vegetation are conspicuous features in the fen basins of the great limestone plain of central Ireland. It is upon them that the inhabitants of the central plain have depended for their peat fuel through the centuries. Almost every Irish raised bog has had much peat removed from the edges, but the fact that many of the bogs are still extensive gives an idea of the enormous masses of peat that have been formed. A few raised bogs still exist in northern and western England, Wales and Scotland, but most of the British raised bogs have long ago disappeared as the result of draining and peat cutting.

Blanket and raised bogs are usually colonized by the ling (*Calluna vulgaris*), but ling does not become dominant while the bog remains wet. If, however, the bog is drained and dries out it usually becomes covered with heath vegetation—often pure Callunetum—and may be invaded by birch or pine, so that birchwood or pinewood is sometimes developed on the dried-out bog peat.

Both blanket and raised bogs are commonly known as 'mosses' in the north of England and in southern Scotland whether or not they are dominated by bog moss. 'Featherbed Moss' is one characteristic name of a plateau 'moss', but most 'mosses' are called after localities. In some bogs *Sphagnum* is comparatively infrequent, and they often consist mainly of great stretches of cottongrass as on the Pennine plateaux, or of deer sedge as in the north-west Highlands.

On many gently sloping areas where the soil is not continuously saturated with water the bog dominant is often mixed with bilberry (*Vaccinium myrtillus*) and ling (*Calluna vulgaris*). Such areas are transitional between bog or 'moss' and heath, while on steeper, better-drained slopes one of these plants is dominant, sometimes the two together, and we have true upland heath.

On peaty soils 6 or 8 inches thick (15–20 cm) overlying siliceous rock, and on the peat debris eroded from the edges of the peat plateaux, the mat grass (*Nardus stricta*) is dominant. On still thinner peaty soils the wavy hair grass (*Deschampsia flexuosa*) often takes its place. A third species, the purple moor grass (*Molinia caerulea*), occurs where the peaty soil is well aerated by percolating (moving) water, and also where it obtains a greater supply of bases. *Molinia caerulea* frequently occurs also on the damper parts of lowland heaths, and on the edges of 'mosses' as well as on fens.

All this last-named vegetation is characteristic of the country which is commonly called moorland, and is associated with wide stretches of heather or bilberry—the upland heaths. Such are the typical 'grouse moors' of Yorkshire and Scotland, good examples of Calluneta. Moorland is distinct from the much wetter bog or 'moss', where the vegetation is on thick wet peat, but the transitional areas in which ling and bilberry are mixed with cottongrass are also called 'moors'. The type of grassland (pasture) dominated by *Agrostis tenuis* and *Festuca ovina*, mentioned on p. 61, also shows transition to heath (both upland and lowland) by the coming in of species such as *Deschampsia flexuosa* on thin peat or peaty humus, and of the heath plants proper where pasturing is reduced or abandoned. On the slopes of the southern Pennines there is fluctuation between this grassland and upland heath according to the amount of pasturing on the one hand, and on the other the tendency to form peaty humus which is a constant factor in the moist climate (Adamson 1918).

The wet climate of the hill country of the north and west of Great Britain leads not only to the leaching of the surface layers of soil owing to heavy rainfall, but also to the building up of acidic humus below the herbage owing to the even more important climatic factor of almost constantly moist air combined with low temperatures, which impedes the natural process of humus decay. The result is seen in the prevalence of heath and moor plants in the semi-natural pastures even on well-drained hillsides and even above limestone rock, except where 'flushes' from springs or from surface drainage continually bring down fresh supplies of bases dissolved from the rocks.

With the exception of the higher-lying and most exposed areas, all these plants communities now occupy country which probably at one time bore forest—at lower altitudes Quercetum petraeae, at higher altitudes Betuletum pubescentis, and at one time Pinetum sylvestris. Felling and grazing have been largely responsible for the disappearance of oakwood and its replacement by heath and grassland, so leading to the general bareness of the hilly regions of the north. Pine rarely spreads from plantations as it does on the drier southern heaths, possibly because of browsing by deer, or because of the soil conditions and the slow seedling growth among deep moss.

The birch and pinewoods whose remains are preserved in the peat may be looked upon as the southern extensions, at the higher altitudes, of the birch–pine association of north-central and northern Scotland. A fringe of birchwood still exists in many places above the oakwoods of northern England and southern and central Scotland, but there is hardly any certainly native pine except in Scotland.

ARCTIC-ALPINE VEGETATION

While the rounded tops and summit plateaux of the hills whose altitude is round about 2000 feet (600 m) are mainly occupied by moor or 'moss' communities, many of the higher mountains, especially in the Highlands of Scotland, but to some extent also in North Wales, the Lake District, and Ireland, show a distinct vegetation generally known as 'arctic-alpine'. This name is appropriate because the characteristic species are at home in arctic Europe, and some occur also in the European Alps. Arctic-alpine vegetation is found most highly developed and forming varied plant communities on the mountains formed of basic rocks. The summits, ledges and screes of acidic rocks show a few species, but these mountains are generally mainly covered with moor or moss, except where there is bare rock exposed.

All the arctic-alpine *vegetation* lies above the zone of former woodland, though individual *species* may be carried down streams, appearing at lower levels; and some, especially on the Atlantic coasts of Ireland and Scotland, occur at sea-level.

The lowest zone of this vegetation is the so-called arctic-alpine grassland, represented mainly by an association dominated by the viviparous fescue (*Festuca vivipara*) and the alpine lady's mantle (*Alchemilla alpina*), with a number of other characteristic species associated. Above the slopes which bear this type of grassland we come to plant communities mainly composed of lichens and mosses, which are by far the most abundant plants of the arctic-alpine vegetation, including many lowland forms, but some confined to the mountains. Here and there we have the sheltered rock faces, ledges and stream-sides which form the habitats of the greater number of the characteristic arctic-alpine species of flowering plants. The individual communities of this varied vegetation are generally small, owing to the very uneven nature of the ground. Finally, we have the summit plateaux with surface composed of loose rocks ('mountain-top detritus'), which are a very regular feature of the higher British mountains; and this is occupied by a sparse vegetation of which the shaggy moss *Rhacomitrium lanuginosum* is the most prominent member, often overwhelmingly dominant.

MARITIME VEGETATION

The last natural types of vegetation which we have to mention are the maritime communities developed on the sea coast, whose habitats are determined by 'maritime factors'. On the one hand we have the vegetation of blown sea sand forming the well-known coastal dunes (and next to this we may put the closely allied though distinct vegetation of shingle beaches): on the other is the very different vegetation of the mud flats or salt marshes, which are covered by the higher tides.

The dominant factor of the sand dunes is the loose, moving sand, not the sea-salt, for we find communities of very similar type on blown sand on the shores of freshwater lakes, as for instance Lake Michigan in North America. The salt-marsh plants, on the contrary, are primarily determined by their periodic immersion in salt water, and by the fact that they are rooted in soil where the water is always more or less saline. Many of these *halophytes* have a peculiar economy, showing distinctive physiological and morphological features; many of them have succulent leaves.

Dune Succession

The sandy shore seaward of the dunes, which is wetted only by the highest spring tides and by spray, is inhabited, where wave erosion is not too great, by a characteristic open community. The sea rocket (*Cakile maritima*), the saltwort (*Salsola kali*), the sea sandwort (*Honkenya peploides*), species of orache (*Atriplex*) and the sand couch-grass (*Agropyron junceiforme*) are the most prominent members of this 'strand' vegetation. All of these can withstand a certain amount of immersion in salt water, showing a corresponding tendency to succulence; and all in some degree (the last-named especially, owing to its habit of growth) arrest the dry sand blown on to them and form low sand hills or dunes through which the plants grow. The low dunes ('foredunes') formed by *Agropyron junceiforme* may eventually reach a height beyond the range of the spring tides, and this species thus comes to dominate an independent community.

The marram grass (*Ammophila arenaria*) cannot stand prolonged immersion in sea water, but its seedlings fairly freely colonize the sand just out of reach of the highest tides. The plants have an even greater capacity than *Agropyron junceiforme* of pushing up when covered with sand; they grow much more extensively and most vigorously under conditions of substantial sand accretion, so that they form much higher dunes, which they bind and consolidate by the ramifications of their rhizomes and roots. Marram grass clothing the dunes acts as a check to the movement of sand particles, and after a time other species, which cannot colonize moving sand, establish themselves between the plants of the marram. Some of these plants are almost confined to sand dunes, but most of them are also found in inland habitats. Under drought conditions the plants of the high dunes, particularly those with shallow root systems, may become very short of water, but the dense vegetation of the low areas between the dunes—the 'slacks'—usually has a plentiful water supply. This is because the sandy floors of the 'slacks' are originally generally stabilized by water so that the water table lies near to the sand surface. Bare areas in which the sand is stabilized by water are quickly colonized by such plants as creeping bent-grass (*Agrostis stolonifera*), rushes (*Juncus articulatus, J. bufonius*) and creeping willow (*Salix repens*) which may rapidly form a closed vegetation (Plate 10). The sand sedge (*Carex arenaria*) often extends, by vegetative spread of its rhizome system, into the slacks from the drier surrounding areas.

On partially stabilized dune surfaces protected from the most violent winds, mosses such as *Tortula ruraliformis*, and lichens, especially species of *Cladonia*, establish and often become dominant. Various grasses, among which *Festuca rubra* is usually prominent, enter and often largely dominate the fixed dune surface. Old grass-covered dunes may form a poor pasture, and before the advent of myxomatosis were often extensively used as rabbit warrens; they are the traditional sites of golf links. Scrub is frequently developed upon them. The sea buckthorn (*Hippophaë rhamnoides*) is a characteristic native shrub of some of the east coast dunes, and is very often planted. It is not found wild inland (but occasionally naturalized) in Great Britain, though it grows abundantly on alluvial gravels in other parts of Europe. Typical heath is developed on the inland parts of some British dunes, although others are too calcareous, even after leaching for many years, to give rise to heath vegetation. Climax forest develops on old dunes in many parts of the world, but not in the poorly wooded British Isles, probably partly because seed is not available in sufficient quantity. Various trees, however, notably species of pine, are successfully planted on dune soil.

Shingle Beach Vegetation

Shingle beach vegetation has considerable affinity with that of dunes, many species occurring in both. The smaller shingle is often much mixed with sand so that the habitats are similar. Shingle is piled above the ordinary high-water level of spring tides by exceptional tides and on a well-developed shingle beach there is generally a distinct 'storm-crest' marking the last high tide effective in this piling-up process. It is on the relatively immobile shingle beyond this crest that vegetation carried down the landward side by the overwash establishes itself, most luxuriantly and characteristically on shingle 'spits' rather than on 'fringing beaches' which are formed on the edge of the coast and are continuous with the land along their whole length. The plants of shingle beaches are able to grow successfully only in the presence of some fine material between the stones of the shingle; the fine material is usually sand, silt or the humus of sea drift—often cast-up seaweed—which is carried down among the stones.

The vegetation of shingle beaches which are not sandy is characterized by various species of lichens which cover the surface of the stones, and by societies of certain species of flowering plants, among which the sea pea (*Lathyrus japonicus*), the curled dock (*Rumex crispus*), the sea campion (*Silene maritima*), the horned poppy (*Glaucium flavum*)—the last three also on fixed sand dunes—the purple herb-robert (*Geranium purpureum*), and forms of the red fescue (*Festuca rubra*) are characteristic. On old shingle beaches, removed from the direct influence of the sea, an inland vegetation develops, and very often a scrub.

Salt-Marsh Vegetation

Salt marshes, as already mentioned, bear a characteristic halophilous vegetation in which almost all the species are peculiar to this habitat. The mud (or sand) covered by every tide is too mobile for plant colonization:

it is only the stretches which lie above the high neap tides that are ordinarily covered with vegetation. The 'eel-grass' (*Zostera*), however, which is well suited to live submerged for most of its life, is frequently found occupying the flats covered by the neaps, where the race of the tide is not too strong.

The annual glasswort or marsh samphire (*Salicornia europaea* and allied species) is frequently the pioneer on the mud flats, the surface of the mud being often prepared for it by the green alga *Rhizoclonium*. The glasswort is frequently followed by the sea meadow grass (*Puccinellia maritima*). The course of the succession is now very variable according to local conditions, but taken as a whole a mixed vegetation (the general salt-marsh community) appears on the flats covered by most of the springs. In this the sea aster (*Aster tripolium*), sea plantain (*Plantago maritima*), sea arrow grass (*Triglochin maritima*), sea spurrey (*Spergularia media*) and sea blite (*Suaeda maritima*) are general constituents. Two prominent plants, which often become dominant about this level, are sea lavender (*Limonium vulgare*) and thrift or sea pink (*Armeria maritima*).

In many parts of the south and south-east coast of Britain, however, the pioneer colonizer of muddy flats is often now the hybrid cord-grass, *Spartina × townsendii*, rather than species of *Salicornia*, and a somewhat different succession operates. The vigorously growing *Spartina* is very efficient in trapping silt and is a strong competitor of many of the characteristic salt-marsh plants, such as *Puccinellia maritima* (Plate 11). In many localities *Spartina* forms a nearly pure stand, the general salt-marsh community being often much restricted to the inland parts of the marshes.

The salt flats covered only by the higher spring tides bear a turf which is a direct result of the closing and consolidation of the general salt-marsh community. This is nearly always grazed and is good sheep and cattle pasture, or 'saltings', the French *prés salés*. The turf is composed of *Puccinellia maritima* or a form of *Festuca rubra*, with *Suaeda maritima*, *Armeria maritima*, *Limonium vulgare* and others. Locally on the south and east coasts of England the shrubby *Suaeda fruticosa* occupies a rather distinctive habitat on the edge of shingle beach where this overlies salt marsh. Other plants which occur in the various special habitats at the higher levels of the salt marsh are the sea purslane (*Halimione portulacoides*), the sea wormwood (*Artemisia maritima*), the stag's horn plantain (*Plantago coronopus*), the mud rush (*Juncus gerardii*), and locally several species of sea lavender (*Limonium*).

The highest levels of the salt marsh, reached by only the very highest spring tides—perhaps only by a few tides twice a year—is often marked by a belt of the sea rush (*Juncus maritimus*), and at the same level also commonly present are such plants as sea milkwort (*Glaux maritima*), sea couch-grass (*Agropyron pungens*) and sometimes other species such as parsley water dropwort (*Oenanthe lachenalii*) which is also found in brackish marshes. These species do not show the characteristic structural features of halophytes to nearly so great a degree as those occurring at lower levels; mixed with them are ordinary land species, sometimes called glycophytes, which will tolerate only small amounts of salt in the soil.

When the tide is effectively kept out of a salt-marsh area and there is adequate drainage by ditches and sluices, the salt is rapidly washed out of the soil and non-halophilous species colonize the area. Excellent pasture-land, which can be converted into arable, is regularly formed in this way. There is no good evidence that salt marsh can develop by the mere accumulation of silt or humus, without human assistance, into a non-maritime vegetation.

VEGETATION OF MAN-MADE HABITATS

As explained in Chapter 2, human activity is constantly providing the most varied new habitats, and these become occupied by communities of plants which may be temporary and casual, but may be closely adjusted to their environment, as are all old communities characteristic of habitats of long standing, whether natural or due to human agency.

The species of 'artificial' habitats, such as spoil heaps, quarries and railway banks, are derived from various sources and include weeds and casuals together with some members of old-established communities able to colonize them.

Spoil Tips

Heaps of deposited mineral or organic material cover a considerable area in Britain. According to its origin, spoil varies widely in chemical composition and conditions for plant growth. Some mine spoil is highly acidic and many remain toxic to plants for many years. However, where spoil is of fertile mineral soil, colonization is extremely rapid, and here the first plants to appear are those with highly efficient methods of dispersal. Among the genera represented are those which include wind-dispersed annuals (e.g. *Senecio*), perennials (e.g. *Epilobium*) and woody plants (*Betula, Salix*).

Where the fertility and the stability of the soil permit, a complete cover of herbaceous plants is quickly established and the pioneer annuals are eliminated by competition. The species characteristic of this second phase of colonization, e.g. bracken (*Pteridium aquilinum*), rosebay willow-herb (*Epilobium angustifolium*), are those in which efficient dispersal is allied to a growth form (tall stature, much-branched underground system) which facilitates rapid and intensive exploitation of both soil and aerial environment. Succession may ultimately progress to scrub and woodland.

Where spoil is favourable for plant growth as regards mineral nutrients, colonization is often checked by erosion. In these situations, species with extensive, often deeply penetrating, rhizomes, e.g. creeping thistle (*Cirsium arvense*), coltsfoot (*Tussilago farfara*), bindweed (*Convolvulus arvensis*) and horsetail (*Equisetum arvense*) may be successful. These species are also prominent where mineral soil is overlain by coarse spoil, e.g. brick rubble on demolition sites or cinders on wasteland.

On mine workings contaminated by toxic concentrations of metals such as zinc, copper and lead, little or no plant growth may be possible. The major colonists are grasses, notably common or brown bent (*Agrostis tenuis*) in which genetic races have evolved capable of growing in soil toxic to most plants.

Refuse dumps and rubbish tips may contain a variety of organic and inorganic material and form ephemeral habitats supporting distinctive assemblages of plants often with very effective means of spread (Darlington 1969). Alien species commonly occur as casuals on these tips and some may become well-established alongside British native species as succession proceeds in the less disturbed sites.

Many species have increased in abundance since the practice of depositing spoil has become widespread. Some of these are most commonly found on one particular type of spoil (Table 6.1).

Table 6.1 Examples of species associated with particular types of spoil

Spoil type	Species
Refractory sands	*Lycopodium* spp.
Cinders	*Chaenorhinum minus*
Toxic (Cu, Pb, Zn, etc.) mine spoil (acidic)	*Agrostis tenuis*
Toxic (Cu, Pb, Zn, etc.) mine spoil (calcareous)	*Minuartia verna, Thlaspi alpestre*
Mortar rubble	*Senecio squalidus*
Farmyard manure and sewage residues	*Urtica dioica, Chenopodium rubrum, Rorippa islandica*
Refuse tips	*Chenopodium rubrum, Matricaria matricarioides, Phalaris canariensis*

Derelict Land

Man has also had a great influence on the vegetation of sites which he has previously exploited but which are now no longer subject to any form of deliberate management. Dereliction may arise from abandonment of farmland, particularly obvious where hill pastures have been neglected in favour of intensive methods of farming in the lowlands. In country districts, the major areas of derelict land are railway banks, quarry margins and unmanaged commons, whilst in the town wasteland includes demolition sites and unattended ground adjoining factories.

As in the majority of terrestrial habitats, the vegetation of wasteland tends to undergo succession towards woodland. This process is well illustrated by the changes which occur when a pasture is abandoned. The most obvious initial change is the replacement of the sward by tall coarse vegetation and there is a marked reduction in the number of species per unit area of ground. In the absence of grazing, tall plants such as false oat (*Arrhenatherum elatius*), couch-grass (*Agropyron repens*), stinging nettle (*Urtica dioica*) and bracken (*Pteridium aquilinum*) grow unchecked and often one of these dominates to the virtual exclusion of other herbaceous plants. As already mentioned, in grasslands over chalk or limestone the widespread removal of sheep and the reduction in the numbers of rabbits following myxomatosis has resulted in an increase in coarse grasses, particularly of tor-grass (*Brachypodium pinnatum*), a strongly rhizomatous species which produces a thick layer of persistent litter on the ground surface (Wells 1969), so suppressing smaller herbs. Where dereliction occurs in moist situations, the great hairy willow-herb (*Epilobium hirsutum*) and butterbur (*Petasites hybridus*) frequently dominate. On fertile soils the tall herbaceous plants may become overgrown in summer by scramblers and climbers such as the larger bindweed (*Calystegia sepium*), goosegrass (*Galium aparine*) and wild hop (*Humulus lupulus*). Where fertility is low, as for example on shallow or highly acidic soils, the response of the vegetation to abandonment may be slow and occasional management by mowing or burning may be sufficient to arrest succession.

After several years of neglect, woody species are conspicuous in most derelict sites. Often these have originated as saplings of shrubs such as hazel (*Corylus avellana*) or hawthorn (*Crataegus monogyna*) which were present in the original herbaceous vegetation but suppressed, often over many years, by grazing, mowing or burning.

Finally, provided that management is not reimposed, derelict sites may progress to woodland. Such woodlands differ from their ancient counterparts in having a low diversity of trees and herbaceous plants. On acidic soils the ratio of birch (*Betula* spp.) to oak (*Quercus* spp.) is usually higher than in natural woodlands and on basic soils the introduced sycamore (*Acer pseudoplatanus*) frequently accompanies ash (*Fraxinus excelsior*) as a co-dominant, or may dominate locally.

Only a very brief outline of the nature and general relationships of British vegetation is given in this chapter. Some plant communities of small extent, for example sea cliffs and 'flushes', developed in areas irrigated by water from springs or from drainage, have necessary been omitted; so also have some man-made habitats such as roadways and paths which bear plants such as pearlworts (*Sagina* spp.) and greater plantain (*Plantago major*) tolerant of compaction and treading. Nevertheless each and every habitat, natural and artificial, presents a wealth of biological problems and provides good material for close ecological study.

Part III

METHODS OF STUDYING VEGETATION

Chapter 7

Scope and Aims of Ecological Work

As environmental biology is of very wide scope it is even more important here than in other scientific fields to consider the aims which should be kept in mind in undertaking ecological investigation. There is no reason why the work of every observant student should not add to general knowledge as well as to his own, for there is still much about our vegetation which remains undescribed. But if the work is to be of value it is essential to know exactly what is being aimed at and to use methods appropriate to the specific objectives; the student should moreover be prepared to revise his aims as the research proceeds if such a change proves desirable.

Every kind of scientific investigation has two stages—the descriptive and the analytical. We must first know clearly what the phenomena are— the features or situations or processes which we propose to study; and we must therefore carefully observe and accurately record before we can proceed to find out how particular phenomena arise. Unintelligent description of badly selected things is worthless but intelligent description of the right features, undertaken in the proper way and with a definite objective, is not only valuable but indispensable to the progress of science.

We should, however, never be content with *mere* description if it is possible to go on to an investigation of causation—of *how things come to happen in the way they do*—for that is the ultimate aim of science. Nor is it necessary or desirable to divide our work sharply into two parts, descriptive and analytical, and to finish one before beginning the other. The careful description of some sets of phenomena, for instance of some natural processes, especially if the description has a quantitative basis, may at once

75

enable us to understand their causes, which we could never have done without it. Even if the problem cannot be solved immediately, a careful description will enable us to formulate it more clearly and precisely, so that we can proceed to the necessary analysis or to experimental work in the field or the laboratory.

The scientific description of types of British plant communities is now fairly complete, a good beginning being made during the first decade of the present century. In addition to descriptive accounts of vegetation, however, it is desirable to have a framework into which the wide variety of vegetation can be fitted, so that this can be more readily understood; it is therefore necessary to classify vegetation. Classification of vegetation may be based on a series of plant communities, representing both climax and seral vegetation, recognized as corresponding with, for example, different types of habitats and soil conditions. Much progress has been made in Britain in this direction, but in addition attention has been directed towards special ecological problems and in recent years investigations have become more experimental. We are now concerned, for example, with the analysis of factors concerned with vegetational stability and change; and much experimental work is being undertaken which bears on the existence and performance of plants in relation to environmental conditions, especially in the particular habitats of the plants. Much valuable descriptive knowledge is assembled in the comprehensive work *The British Islands and Their Vegetation* (Tansley 1953) first published in 1939, and likely to remain for long a standard reference to British vegetation. Somewhat different descriptive procedures and particularly of means of classifying vegetation have, however, been adopted in phytosociological study on the continent, in which stands of vegetation are categorized by particular assemblages of characteristic species. The fundamental unit of vegetation is regarded as the Association (used here in the sense of Braun-Blanquet (1932) and not identical with the association referred to on p. 30 and elsewhere), and is characterized in part by species of high constancy (species occurring in a high proportion of the stands) and also 'fidelity' (species which are completely or partly restricted to a particular Association). A hierarchical system of classification is adopted in which Associations (names ending in -*etum*) are grouped into Alliances (ending in -*ion*), Alliances into Orders (ending in -*etalia*) and Orders into Classes (ending in -*etea*). Methods similar to these of the Zurich–Montpellier school, which are widely used by ecologists on the continent, have now been applied to British vegetation, especially to that of Scotland, and these procedures are being increasingly adopted. Such studies, which permit a close comparison of British communities with those distinguished on the continent, form the basis of the extensive account of *Plant Communities of the Scottish Highlands* by McVean and Ratcliffe in 1962 and are widely drawn on in the comprehensive descriptions in *The Vegetation of Scotland* (Burnett 1964). For further discussion of the Braun-Blanquet system reference may be made to the appraisal by Poore (1955*a*, *b*). Details of the methods of the Zurich–Montpellier phytosociological school, the historical development of this school, its relation to other systems of classification of vegetation, and also a synopsis of the units of vegetation present in Britain according to the Zurich–Montpellier classification are given by Shimwell (1971).

In Britain, the method of primary survey (see Chapter 8), on which the general classification of associations rests, was to choose a stretch of country and then distinguish and map the communities that could be conveniently represented on the scale of 1 inch to the mile. The first surveyors had, of course, everything to learn. They had to determine the kinds of vegetation which they would select for representation; and later surveyors, besides finding new types in new areas, did not always agree that the original ones were rightly chosen.

The usefulness of primary survey nevertheless remains, and the symbolic representation of the components of vegetation in black and white, though not as effective as colour used in some early maps, is satisfactory. There is no better training in extensive study of vegetation than carrying through the primary survey of a suitable area, because many different types are encountered in the course of the work and the relations of these have to be closely studied to represent them properly on a map. The making of a good map—although not to be regarded as the ultimate goal of any ecological work, which is the understanding of the vegetation—is a most useful exercise in that it focuses effort and compels the student to make up his mind about the status of the vegetation. As seen in the next chapter, it is often necessary to 'interpret' vegetation, since it is impossible to record every plant population on a small scale exactly as it is, and making such interpretation is in itself a valuable training.

There is, however, another method of extensive work, which leads perhaps to a deeper understanding of individual plant communities. This may be called the *monographic method*. It consists of taking a single well-defined plant community, a consociation or an association, such for instance as the oakwood, the heath or the reedswamp, and following it wherever it is to be found throughout the country, studying the variations of its composition and structure, its dependence upon habitat factors, climatic, edaphic and biotic, and its relations to other communities into which it passes by gradual transitions.

An ideal study of this sort should be extended not only throughout the country but over the whole area of its occurrence in the world; but where this is not possible the monographic method applied within narrower limits will give a deeper insight into the community selected than can be obtained during a general survey of the vegetation of a restricted area.

Both the primary survey and the monographic methods are essentially *extensive*, carried on over a wide area of country primarily by the methods of observation and comparison. On the other hand, what may be called *intensive* work is concerned rather with detailed investigation of particular problems in one spot. These problems are endless and may be pursued to any length that the inclination, opportunities and means at the disposal of the student may dictate. Among them may be mentioned the whole set of problems centring round the *multiplication* and *dispersal* of individual species: the amount of fertile seed a given species produces under different conditions, the means by which the species spreads and its rate of spread; the conditions under which germination can occur and establishment of the seedlings take place. Closely linked with these problems are those concerned with *competition* between individual plants, either of the same or of different species: the success or failure of one or the other of two competing species and the way in which this is brought about, for instance, by root competition or overshadowing; the effect upon the result of different conditions of the habitat, for instance, the moisture content or other properties of the soil. Then there are the one-sided or mutual benefits which plants may confer upon one another: for instance the protection afforded by a larger plant by way of shelter from desiccation, or by a spiny plant in keeping off the attacks of browsing animals; and again the preparation of the soil by the formation of humus from the decaying parts of one plant for the establishment of others, and so on. On the solution of numerous problems of this nature depends the detailed understanding of every case of natural *succession* (see Chapter 4), and at least some of these can be studied with every prospect of success by anyone who lives in the country and has time at his disposal.

In the study of such problems it is important to employ experiment as well as observation, wherever possible. This can be profitably done in the field wherever there is little or no risk of disturbance. The likelihood of interference is a serious drawback to the setting-up of experiments on vegetation in the English countryside, but it is often possible to obtain the tolerance or even the interest of landowner, farmer or keeper, and thus help to secure the safety of field experiments. And some experiments which do not require apparatus of any sort are quite inconspicuous. Exact positions of such experiments can usually be successfully marked by wooden pegs driven in almost flush with the soil surface as long as the experimental area can be approximately located by reference to some conspicuous and persistent feature of the habitat.

The importance of field experiments in ecology cannot be overestimated. It must not be supposed that experiments necessarily imply instruments or apparatus, elaborate or simple. The experimental method simply means the observation of processes under controlled conditions. The nature and extent of the control necessary, are, of course extremely varied, according to the problem under investigation. For instance, the fencing of a small area against rabbits (Farrow 1916; Tansley 1922; Watt 1923), the cutting of small, narrow trenches to divert the water supply which ordinarily flows down a slope in a region of heavy rainfall (Jeffreys 1917), the supply of extra water to vegetation growing on dry soil (Farrow 1917b), the addition of a range of mineral nutrients to different types of vegetation (Willis and Yemm 1961; Willis 1963, 1965)—all these are simple field experiments, many of them of long-standing, which have given valuable results. Such experiments are essential for the solution of many ecological problems, and moderate ingenuity will suggest numerous ways of testing the effects of the different factors at work upon vegetation.

Many ecological processes can be conveniently studied in a garden, but the conditions in a garden are, of course, very different in many respects from those of natural vegetation, and caution is required in making inferences from one to the other. With this proviso a great deal of information can be obtained from garden experiments. Natural conditions can sometimes be closely imitated by growing plants in competition in boxes filled with the natural soil of their habitats. But such work can never wholly replace observation and experiment in the field.

Something may be said here about quantitative methods. The determination of exact quantitative results from which quantitative laws can be formulated is a well-established method of science. Broadly, we may say that in proportion to the advance of a branch of science its methods become quantitative. This is as true of biology in general and of ecology in particular as of other branches of science. But neither biology in general nor ecology in particular can as yet be treated wholly by quantitative methods, though these should be applied wherever it is appropriate. Nevertheless, in recent decades there has been extensive development of quantitative procedures for investigating the structure of plant communities, outlined in Chapter 11, and statistical techniques have been applied to the analysis of the relative importance of the influence of various environmental factors on plants. By increasing use of such techniques ecological investigation is now becoming a more exact science.

However, we must never make a fetish of quantitative data, as they are not *necessarily* valuable. To be of value for making inferences and ultimately formulating 'laws' they must have some kind of *general validity*, forming parts of a general description or being causally related to one another. Isolated data are of little or no value. This is equally true of enumerations of species and frequencies, and of quantitative records of the factors of the habitat, such as soil analyses, evaporimeter readings and light records. The mere taking of an instrument into the field and making readings, just as the mere recording of the number of individuals of a species in a given area, is not a virtue in itself and is no guarantee of scientific results. In seeking causation there is always a definite object in view, a definite problem to be borne in mind, and observations and experiments must always be directed towards solving it.

On the other hand, intelligent *qualitative* observation of the constitution and relationships of plant communities sometimes at once enables us to recognize causes, or at least clearly to state problems for solution. In so far as qualitative observation contributes to a picture of the vegetation of a country or region, it is essential in the first descriptive stage of investigation. Quantitative data may be added by way of closer characterization, but it is of little use to multiply these indefinitely until specific problems arise and can be clearly stated and deliberately attacked.

In attempting to solve a specific problem the best mode of approach should first be carefully considered. Ecological problems are usually complex, and the first approach may not give the desired results. One problem leads to another, and it may be necessary to abandon the attempt to solve the first until the second has been stated and explored. The attempt to solve the second may, in its turn, show that a third must first be attacked, and a successful solution of that may cause the first to disappear altogether. For instance, suppose we are trying to explain the puzzling distribution of two or more plant communities over a certain area of ground. At first it may be thought that the undoubted variations in texture and chemical constitution of the soil are responsible. But soil analyses may show that none of these variations can be connected directly with the differences of vegetation. There may appear evidence that these differences correspond, to some extent, with the moisture content of the soil, though the communities inhabiting, on the whole, the drier and damper areas respectively are apparently quite capable of growing on either. This suggests a difference in the incidence of competition between the plants of the two soils, and the difference in competition finally turns out to be determined by the attacks of animals which avoid very wet ground, and bear more severely on one of the communities which they prefer, thus handicapping it in its struggle with the other, except on the wet ground where it is not so heavily attacked. In the attempt to determine the observed distribution the original problem, the correlation with soil constitution, has vanished and the question of moisture has become subordinate to that of competition as affected by animal attack, which thus proves to be the main cause of the distribution observed. Such a case as this (cf. early studies by Farrow 1916, 1917a, b, c) clearly brings out the necessity of keeping an open mind, and not persisting in a line of investigation which is not giving good results.

Also the investigator should never allow himself to become enslaved by his *methods*. The methods which seem most suitable should be carefully thought out beforehand, and strictly adhered to until thoroughly tested. But they should be modified or abandoned directly they are proved unsatisfactory and better ones can be substituted. Never adhere to routine for the sake of routine. This warning applies particularly to listing, charting and mapping methods (see Chapters 8, 9 and 10). A great deal of valuable time

4 Birchwood near the Pass of Killiecrankie, Perthshire. The ground or field layer developed in the light shade cast by birch is rich in grasses.

August, M. C. F. Proctor

5 Chalk hills (Knocking Hoe National Nature Reserve) on the Bedfordshire/Hertfordshire border showing impinging agricultural pressures, grazed grassland and an enclosure for the study of grassland management. The straight line on the hill (Hoe) marks the boundary of an area ploughed about 1953; virgin chalk grassland lies to the left, and 'reverted' grassland of little floristic interest to the right.

I. H. Rorison, 1970

6 Pond in Wicken Fen, Cambridgeshire, with water lily (*Nymphaea alba*) and strong growth of the common reed (*Phragmites communis*).

M. C. F. Proctor

7 Fen vegetation in Perthshire. Prominent among the rank growth of herbaceous plants are *Angelica sylvestris* and the meadow-sweet (*Filipendula ulmaria*); birch (*Betula*) trees fringe the fen in the background.

August, M. C. F. Proctor

may be wasted, for instance, by adhering to laboriously accurate methods which are giving little information likely to be of value. This is not to say that detailed charting, for example, in which the position of every individual plant is shown, is not sometimes of essential value. It all depends on the particular object in view.

Chapter 8

Extensive Studies—Reconnaissance and Primary Survey

It is certainly a sound principle that an extensive study or survey of a fairly large tract of country—primary survey as it is called—forms the best preliminary to intensive work. Such a survey gives a general knowledge of the types of vegetation and the conditions in which they occur, and enables the choice of areas or communities for more detailed study to be made with skill and judgment. An approach of this kind is desirable where the vegetation is unknown or little known, but for many parts of Britain substantial knowledge of the vegetation already exists and can be profitably used as a basis for further work. In such instances, primary survey, and its preliminary—reconnaissance—are unnecessary.

RECONNAISSANCE

In relatively unknown country the more cursory work of reconnaissance is a valuable preliminary to primary survey. In reconnaissance the country is rapidly traversed, the salient features of the vegetation noted and occasional samples listed, so that a general idea is gained of what the area is like.

First a good contoured topographical map of the district to be reconnoitred is required. Maps of the Ordnance Survey on the scale of 1 inch to a mile (1 : 63,360) may be used, or Bartholomew's 'layer contour' maps on a scale of $\frac{1}{2}$ inch to a mile (1 : 126,720). In addition, maps which show the geology and soil features of the area are valuable. If possible, soil survey maps, on the scale of 1 inch to the mile published by the Soil Survey of Great Britain should be obtained (see Appendix), but not all the country has yet been surveyed. 'Drift maps' of the Geological Survey on the same scale, which show the surface geology—usually all that matters to the ecologist—although available for only parts of the country, are also useful. From the topographical and geological maps the general nature of the ground and the situation, for example, of woodland and open country, can be seen and the best access routes chosen.

Each well-marked type of natural or semi-natural vegetation seen should be rapidly examined, and the dominant and abundant species, as well as any striking ones, recorded. At the same time the principal agricultural crops should be noted. It is useful to record on the map itself (e.g. the folded maps of the 1-inch Ordnance Survey or Bartholomew's $\frac{1}{2}$-inch maps) the general type of natural vegetation, plantation, arable or pasture land. This can be done by means of symbols, or letters denoting the generic names of the dominants of natural vegetation, or in the case of crops the English names.

Although the only essentials are a map, notebook, sharp eyes and an alert mind, several items of equipment are useful on reconnaissance studies. Of particular value are a range of polythene bags for carrying home plants or soil, a camera, a trowel, and a tightly stoppered bottle of dilute (20 per cent) hydrochloric acid (care is needed because it is corrosive). By addition of this acid to a sample of the soil an indication of the amount of 'lime' (calcium carbonate) present can be quickly obtained from the vigour of bubbling of the carbon dioxide liberated. A rough subjective scale of reaction is: L_0, no bubbles; L_1, few isolated bubbles; L_2, slight general effervescence; L_3, moderate; L_4, strong; L_5, violent bubbling. It should be remembered that the surface layers (the top 5 cm) of the soil often contain almost no lime because of leaching and humus formation, while slightly deeper layers, reached by the roots of some of the plants, may contain a considerable amount.

Successful reconnaissance work naturally presupposes some knowledge of species and also of rocks and the soils they produce and the beginner will need to devote time and attention to identification and recognition. Little ecological study can be done without a good knowledge of the different species which make up the vegetation; every effort should therefore be made to become familiar with plants and to be able to recognize them not only when they are flowering but also when vegetative and at the seedling stage.

No attempt should be made in reconnaissance to cover every part of the area, as in primary survey, or to do detailed studies of plant communities. The aim is only a general result based on observations including all the main types of country present.

Great assistance in reconnaissance work is provided by photographs taken from the air (see Chapter 13). The development of aerial photography in recent years has been one of the most important advances, and has led, for example, to the recognition of features such as former shore lines and outlines of ancient buildings, not easily seen on the ground. Quick and accurate mapping of topography and vegetation is possible from aerial photographs. While confirmation of the interpretation of aerial photographs should be made on the ground, with experience the certainty of interpretation is such that a great deal of mapping can be done directly.

PRIMARY SURVEY

This name is applied to the general method adopted by the first ecologists (Robert and William G. Smith) who began to study British vegetation systematically at the end of the nineteenth century. It consisted essentially of recognizing and describing the larger vegetation units, making lists of their floristic composition, studying their relationships and the general nature of their habitats, and mapping their distribution. Such maps are usefully compared with the series of coloured geological maps issued by the Geological Survey. The maps give a graphic record of the distribution, not only of the main types of natural and semi-natural vegetation, but also of the principal kinds of cultivated land. This last has to be treated somewhat differently from the native vegetation, since it is impossible to map all the crops, which often change from year to year in accordance with a scheme of rotation.

The first primary surveyors did valuable service, providing a basis for the modern systematic and intensive study of British vegetation. Essentially all the primary survey of vegetation has now been done for Britain, but it is still of value and provides excellent practical training for the ecologist.

A great deal of information about the natural and semi-natural vegetation of Britain, as well as about the crops grown, can be gained from the maps produced in the Land Utilisation Surveys. The first of these surveys was directed by L. Dudley Stamp in the 1930s and the maps (on a scale of 1 inch to a mile) show forest and woodland of different kinds, arable land, meadowland and permanent grass, heaths and moorlands, rough marsh pasture and agriculturally unproductive land. The current (second) Land Utilisation Survey is more detailed, the different types of land-use being mapped by means of colouring, symbols and letters. The land-use maps (see Appendix) of this second survey, so far available for only a fraction of the country, are in eleven colours and on a scale of 1:25,000 (approximately $2\frac{1}{2}$ inches to a mile). Much detail concerning the vegetation is included; for example heath, moorland and roughland (shown in yellow) may be overprinted with F for *Festuca–Agrostis* communities, M for *Molinia caerulea* and N for *Nardus stricta*.

The making of accurate maps is a desirable aim in itself. It is important, however, to be quite clear about the purpose of the map, and about the categories to be recognized in it. The aim may be, for example, to show the general appearance, type or *physiognomy* of the vegetation, to recognize units of natural vegetation, to record the dominant species or the distribution of particular species, to show the present land-use or to record soil types or the geology of an area. All of these maps are of value to the plant ecologist, but each map has its own specific purpose.

The area chosen for a primary survey must generally be determined by accessibility, since much time must necessarily be spent in the field. If a choice is available it will clearly fall on a region with diverse natural and semi-natural vegetation, unless interests are primarily agricultural or related to derelict land. If an area is selected coincident with that on one or more map sheets in preference to a 'natural' region, delimited by some geographical feature, more varied data on an equal surface of land are obtained.

Field Maps

These should be the 1-inch, 2½-inch (1:25,000) and 6-inch Ordnance Survey sheets (for suppliers see Appendix) of the selected area. The 1-inch Seventh Series maps (or the 1-inch Tourist maps, covering certain 'tourist' areas with interesting vegetation) are convenient for relatively uniform country such as mountain and moorland and for purely agricultural land as in much of East Anglia. In varied country where the vegetation changes frequently within a short distance, or where detailed mapping is required, 6-inch (1:10,560) maps are needed.

All modern maps of the Ordnance Survey carry the National Grid. This is a series of squares with sides parallel to and at right angles to the central meridian (longitude 2° W) of the Ordnance Survey mapping. The sides of the grid squares are multiples of the metre, and by means of the grid the location of any point in Great Britain can be accurately given. The position eastwards and northwards, from an arbitrary origin near the Isles of Scilly, of any point can be determined, to a precision dependent on the scale of the map. The 100-kilometre squares of the National Grid are denoted by letters, and these precede the figures of the grid reference. The records of plant distribution in Great Britain given in the *Atlas of the British Flora* (Perring and Walters 1962) refer to areas of 10 kilometres square. It is a good practice to record the National Grid reference of any locality at which ecological records are made, as the spot can then always be pin-pointed later if necessary.

Notebooks and Note-taking in the Field

A pocket notebook or recording pad, e.g. paper between strong card-boards fitted with bulldog clips, is required. To avoid possibility of loss of valuable field records, carbon duplicate notebooks can be used, and the duplicate left at home. Each worker will have his own preferences in regard to note-taking, but the great advantage of following a well-thought-out system is that it tends to completeness of record and ease of reference. One of the commonest experiences is to discover, when writing up notes later, the failure to record important or even essential observations. Every opportunity (such as wet days) should be taken to revise notes and prepare draft descriptions of vegetation seen, so that gaps in observations thus revealed can be made good on the next visit to the site.

It is always essential to make notes *on the spot* of every feature that seems worth recording, with scrupulous emphasis on any fact which runs counter or appears to run counter to a previous conclusion or preconceived opinion. The records of facts should be accompanied by, but clearly distinguished from, any hypotheses or conjectures that occur on the spot, or later, while the observations are fresh in the mind. The importance of full systematic note-taking and constant consultation and revision of the notes cannot be over-emphasized.

Remaining Field Equipment

Map and notebook are the only *essentials* in primary survey, but other equipment is frequently required. Items that may be useful are, however, too numerous to carry conveniently on every field excursion, and a selection must usually be made. It is a mistake to carry too much, because of the load and tendency to distract attention from the problems in hand.

For these reasons it is usually desirable to make observation, recording on the field map, listing and note-taking, the main objects of the first working of a section of country; soil testing, the collection of soil samples, light records, photography and other detailed work can be left for a second visit, on which, of course, the first records can be checked and if necessary revised.

It is always well to carry a *pocket flora*, such as the *Excursion Flora of the British Isles* (Clapham, Tutin and Warburg 1968) for determining doubtful species on the spot. A good *pocket lens* should always be carried by every botanist. The beginner (and sometimes the more advanced student) who does not know his species well will have to take home with him specimens (but not rare plants) for determination. For this, polythene bags of assorted sizes are valuable, being less heavy and less bulky than the vasculum or portable press often used formerly; the plants keep in fresh condition as long as the bags are not crushed and are closed so that the plants do not wilt. Labels, giving the details of the locality of collection, should be put in the bags with the specimens when collected.

A tightly stoppered bottle of *dilute hydrochloric acid*, as already explained, is frequently useful, though not wanted on siliceous soils which are known to contain minimal amounts of calcium carbonate. Often it will be necessary to take measurements, for which a flexible steel tape is convenient. A *compass*, preferably prismatic, is useful and even necessary in wild country. All the above can be carried easily enough in a light haversack, or in roomy pockets, without encumbering the surveyor.

A *camera*, polythene tubes or strong *polythene bags* for soil samples, a strong *trowel* or trenching tool for examination of root systems and for collecting soil samples, and a *soil borer* (auger) are relatively bulky, and are often better left for particular excursions. Details concerning photography and the study of soils are given in Chapters 13 and 15. Here it need only be said that almost any camera may be used for snapshots of general views of brightly lighted vegetation.

Method of Work in the Field

On approaching a particular community it is advisable first to concentrate attention on a small area which appears typical, and to select a spot where the vegetation can be examined in detail, the species present identified, and a short description written. Second and third spots may then be examined by the same procedure. In this way an accurate idea of the species present and of the structure of the community is gradually built up.

One of the first tasks is to decide by what system the vegetation is to be classified and to determine what plant communities are to be recognized for mapping purposes. In Britain these units will be usually and primarily associations and consociations of the natural and semi-natural vegetation. Typical samples must be thoroughly examined, species listed and the main features of their structure recorded. The consociations should then be recorded on the field map by suitable symbols. Boundaries between different consociations or associations should be drawn, and where there are zones of considerable width, transitional between two communities, these should be marked. Sometimes all this can be done on the 1-inch map, but sometimes the $2\frac{1}{2}$-inch or 6-inch map will be required.

Many difficulties arise, largely caused by the modification and fragmentation of the vegetation through human activity. Areas which present too many and bewildering difficulties of interpretation and judgment may be roughly described in the field notebook and then left until further experience has been gained. This will often automatically clear up the problems, and then the doubtful areas can be revisited and included in the map. The ideal of recording the vegetation actually on the ground cannot always be strictly kept. It is impossible to represent on a comparatively small scale, such as 1 inch to a mile, all the variations, whether caused by local differences of habitat or by human interference. Many such variations will have to be grouped under one type. One of the first tasks is to decide upon the types to be recognized but these may have to be modified later. The notebook should of course contain full notes on the variations, and every effort should be made to discover their causes.

The following gives an example of what is meant. Suppose a woodland of definite type extends over a certain tract of country. Most of it may be in a stable semi-natural condition; for instance, it may have standards of a certain species of dominant tree, and coppice of certain species of shrubs in fairly constant proportions, with a ground vegetation of similarly uniform characteristics. Clearly that tract of country must be represented as uniform, and marked with the initial of the dominant tree, even though it may include some areas which differ from the type more or less widely. The ground may be wet in some places because the water-level is close to the surface, and the typical vegetation of the woodland may here be mixed with or replaced by other species which are favoured by wet soil, though such local societies are much too small to show on the map. If a special study of the woodland is being made, they may of course be shown on the 2½-inch or 6-inch map. In some places there may be no standards at all, but only coppice of the same general type as where the standards are present. Other areas may be completely replanted with conifers; if these are large enough, they can be indicated. In others, again, the wood may have been cleared or partly cleared and left derelict, or cattle may have been allowed to graze through it, causing considerable modification of the vegetation such as the entrance of many species alien to the woodland and the disappearance of some woodland species.

The whole area can, however, be mapped as belonging to the typical woodland when it has quite clearly been derived from this by human agency. In process of time, of course, the site of the original woodland may be converted into something totally different, for instance grassland, or it may be ploughed up and turned into arable fields; and then naturally it can no longer be reckoned as belonging to the original woodland. Similar cases are provided by the partial drainage of marshes and bogs, by the heavier pasturing of certain tracts of grassland, and so on. These differential treatments may cause very great differences in the vegetation, but the whole area of marsh or grassland is often rightly treated as a unit for primary survey purposes.

A good deal of useful information may be obtained from conversation with local inhabitants, especially farmers, shepherds or gamekeepers. Experience and observation provide tests which make it possible to sift such information. It is also sometimes possible to get useful data from landowners or their agents about the past treatment of land. Old maps are often very valuable sources of information about the condition of an area in former times.

Listing of Species
The making of careful lists of species occurring in the different communities is a most important task, for the flora of a plant community is its essence. It is often not a straightforward matter to make a satisfactory list, because of the modification of the original composition of the community, especially by the entrance of species from outside owing to human interference. One may often find, for example, in a coppiced wood a number of species which are not true woodland species, but which enter the wood when it is opened up either by the agency of wind or because their fruits or seeds are carried there by traffic (transport on footwear or in clothes of people, particularly men coppicing and felling, in mud on the hooves of cattle and horses and the wheels of vehicles, is important). Such species are able to establish themselves in the wood because of the light which reaches the ground owing to the absence of a continuous thick tree canopy such as exists in 'high forest'. Some are distributed through the coppice,

8   Young carr dominated by alder buckthorn (*Frangula alnus*), with marsh fern (*Thelypteris palustris*) and the sedge *Carex elata* in the field layer, Wicken Fen.

July, M. C. F. Proctor

9   Blanket bog near Slieve League, County Donegal, in which the stumps of pine are exposed by peat cutting and erosion. The peat is partly vegetated by mat grass (*Nardus stricta*), rushes (*Juncus* spp.) and cottongrass (*Eriophorum*).

M. C. F. Proctor

10 Dune vegetation at Braunton Burrows, N. Devon. The dunes are stabilized by marram grass (*Ammophila arenaria*) but the flanks are eroding. The floor of the 'slack' is densely vegetated with creeping willow (*Salix repens*) and bent grass (*Agrostis stolonifera*). In the foreground are isolated patches of *Salix repens* in an area of sand accretion with *Ammophila arenaria* and sand sedge (*Carex arenaria*).

June, M. C. F. Proctor

11 Salt marsh at Budleigh Salterton, Devon. The cord-grass (*Spartina × townsendii*) is strongly established, dominating large areas. Its tall thick growth contrasts with that of the sea meadow grass (*Puccinellia maritima*) which forms a low turf.

May, M. C. F. Proctor

others are confined to rides and pathsides (dispersal may be partly in-volved here, but the two habitats differ in several ways and support differ-ent species and societies). They may have very different origins, but many of them belong to the class often called 'marginal', i.e. they are charac-teristic of the semi-shade of wood margins, hedge banks, etc. Others may be pasture plants or wayside or arable weeds. Similar cases of the existence together of species of very different origins occur in many other modified habitats. Such different categories of plants are often not easily distinguished, however, and the only safe course is to make a complete list of the species within the limits of the community.

When a community is definitely stratified, it is best to list the constituents of the different strata separately. In a wood it is generally quite easy to distinguish the strata (see p. 34), but in some communities, particularly in grassland, the stratification may not be very definite, i.e. the plants may vary greatly in height and some of the species may bridge the space between two successive strata, expanding their basal leaves, for instance, in one stratum, and their upper leaves in a higher one. It is always well to spend some time studying stratification in such a community, for it is an import-ant structural feature of the community and is of great ecological interest because the actual habitat conditions may differ considerably in the differ-ent strata. Well-marked societies in which the flora differs distinctly from that of the rest of the consociation or association should also be listed separately.

When listing is required on a broad scale, and especially where a number of different examples of the same type of community are to be examined, it is convenient to prepare a list comprising all of the common species likely to be encountered, and to use this list as a basis for recording. In this way it is less likely that species will be overlooked, and comparisons between sites can be easily made. In extensive surveys, such as those carried out by the Nature Conservancy for example, a series of printed cards on which the presence of species (and of other standardized information) can be readily recorded prove of value.

Frequency Symbols

An index of the frequency of each species recorded is desirable and should be added to its name. In British lists the following symbols have often been employed:

$$d = \text{dominant} \qquad f = \text{frequent}$$
$$co\text{-}d = \text{co-dominant} \qquad o = \text{occasional}$$
$$va = \text{very abundant} \qquad r = \text{rare}$$
$$a = \text{abundant} \qquad vr = \text{very rare}$$

The letter *l* is prefixed to the symbol when the dominance, abundance or occurrence is *local* only; thus *la* = locally abundant.

Dominance has reference to a given layer only: thus the pedunculate oak (*Quercus robur*) may be dominant in the tree-layer of a wood, the hazel (*Corylus avellana*) in the shrub-layer, the primrose (*Primula vulgaris*) in the herb-layer, and there may be a moss-layer, in which for instance *Atrichum undulatum* or some other species is dominant. When two or more species share the dominance in a given layer, *co-d* is used, but in some communities there is such a mixture that no species or group of species can be said to be generally dominant. This is particularly the case in seral and all transitional communities, but may also apply to some climax communities such as tropical rain forest.

The assignment of frequency symbols depends of course upon a subjective judgment without a fixed quantitative standard, such as could be obtained for instance by recording the species occurring within each of a large number of small areas of uniform size taken at random. Subjective estimates of frequency may vary considerably, not only between different observers but between records made at different times by the same observer, particularly in the categories 'occasional' and 'frequent'. Nevertheless subjective estimation gives useful preliminary data.

It should also be clearly realized that such estimates vary with the size of the area listed. Suppose, for example, that an afternoon is spent traversing in different directions a patch of woodland about a mile (1·6 km) square with a uniform flora, and that the different species are noted as they are encountered, the frequency letters being added from time to time and checked and corrected with further observation. An 'abundant' species will be one which is never or hardly ever out of sight, a 'frequent' species one which is not abundant in this sense, but nevertheless is constantly being found, an 'occasional' species one which is seen perhaps ten to twenty times in the course of the afternoon, while a 'rare' species is seen only once or twice. But if a small area of the wood, say an acre (0·4 ha), is thoroughly searched and the species within it noted separately, the species called 'rare' in the wood as a whole will probably not be seen at all, even the 'occasional' species may not be present, while the species 'frequent' in the wood as a whole may fall to the rank of 'occasional' or even 'rare'. If, on the other hand, the frequencies of species in a large number of woods of the same general type are considered, the plants which are 'rare' or even absent in a particular wood may be appropriately called 'occasional' in a wide stretch of country; if a certain number of specimens occur in most of the woods, the 'occasional' plants of the single wood may become 'frequent', and so on. Thus these terms have a significance which is strictly relative to the size of the area considered, and this fact must always be remembered in compiling lists. The frequency symbols given are most suitable for the larger areas, and when a general account of such an area is being made the number of listed examples of a given association in which they occur, as well as their frequencies within such examples, must be taken into consideration.

Seasonal Changes
In describing and listing an association or consociation it is very necessary to remember the seasonal change in vegetation. This is most marked in deciduous woods, where there is a very distinct 'prevernal' ground vegetation; several species of this ground flora, such as *Adoxa moschatellina*, *Anemone nemorosa*, *Endymion non-scriptus* and *Ranunculus ficaria* may easily be missed altogether if the wood is visited only in late summer, their leafy shoots largely or completely disappearing soon after flowering. But it applies to other communities also. The little spring-flowering annuals ('ephemerals'), such as whitlow grass (*Erophila verna*) and rue-leaved saxifrage (*Saxifraga tridactylites*), of open spots in dry grassland (e.g. fixed sand dunes) completely disappear in the summer. Many of the orchids of chalk grassland are not to be found in late summer or autumn. On the other hand, heath, fen and salt-marsh vegetation do not reach their full development until after midsummer.

Visits should be paid at as many seasons of the year as possible. If only three visits are possible in the year, these should be in April, June and August; if only two, early May and July for woods; if only one, then it should be in late June for woods and grassland (except that which is cut for hay), July or August for heaths and fenland, August or September for salt marsh. These months are suitable for most of England and Wales; in the north and in Scotland, particularly the east, most of the vegetation is from three weeks to a month later in the average year.

Objects of Primary Survey
The main object of an extensive survey is to distinguish, record and characterize the larger communities met with and to note their relations to topography, exposure, soil type and ground water. But special note should be taken of the relations of the different communities to one another, of advance or retreat, and of successional phenomena in general—modified, as they are nearly everywhere in this country, or even caused, by the effects of human activity, pasturing of various kinds, drainage, felling, burning, etc. 'Marginal phenomena' of vegetation, i.e. modifications brought about, or special communities developed, on the limit between well-marked primary communities by the effect of one upon the other, or by the human factor which has differentiated the two, should always be carefully investigated. It is in the course of observations of these kinds that problems which have to be studied by more intensive methods of work can be recognized and noted.

The monographic method (p. 77), in which one plant community, followed over a wide stretch of country—if possible over the whole area of its geographical distribution which will usually involve visits to other countries—is the centre of interest, can be pursued by methods similar to those of primary survey, but the observations will be more detailed. The habits and performance of the dominant and other prominent species of the community will receive special attention; the floristic composition and its variations, the essential features of the habitat and the relationship to other communities will be closely studied. The life history and behaviour of community dominants is a very important ecological study belonging to the field of autecology. Really thorough knowledge of a community cannot be acquired without close study of the autecology of its dominants.

Chapter 9

Surveys of Smaller Areas

Sometimes it is desired to map an area of vegetation, perhaps a quarter of a mile (or half a kilometre) across, on a larger scale, say 1:1000 or 1:500, than can be shown in primary survey on the topographical maps of the Ordnance Survey. Such a map has to be made *de novo* by one of several possible procedures.

The traditional 'chaining' method used in land survey may be employed. A convenient *base line* running the length of the area is first selected. Every part of this should, desirably, be visible from every other part, but this is not possible in some terrain. The two ends are marked with permanent pegs. It is essential to get the base line straight. This may be done by planting ranging poles or rods about 6 feet tall at the two ends and then 'ranging' these with a series of other poles until all are exactly in line. If an observer stands at one pole and looks towards the other end of the line the next pole should completely hide all the others. A second worker goes to each pole in turn and adjusts its position in accordance with hand signals of the observer. When the line is perfectly straight it is carefully 'chained', i.e. measured with a surveyor's chain or tape and permanent pegs driven in at suitable intervals (50 or 100 feet). Tapes made of 'fibron' or other plastic are more convenient and lighter to carry than chains. These plastic tapes do not suffer the disadvantage of tapes formerly used which stretched and shrank, sometimes considerably, when wetted and dried. Tapes of 50 feet and 100 feet long are useful; a further advantage of tapes is that they are available as well in metric units (e.g. 30 m), in which all measurements are now to be preferred.

Once the base line is fixed and measured, perpendiculars are erected on it at suitable intervals, extending to the edges of the area to be mapped. These should be placed where they will run across the greatest number of physical features present and vegetation boundaries. The right-angles can be determined by the traditional methods involving the use of a *cross-staff* or an *optical square*, but can also be obtained by means of two tapes, as shown in Fig. 9.1. If AB is the base line and a right-angle is to be constructed at B, slip the metal loop of a measuring tape on a land survey 'arrow' or marker (a peg or skewer is satisfactory) at B and run out the tape approximately at right-angles in the direction BD. Now slip the loop of another tape on a peg at C, 9 feet from B along BA, and run out in the direction CD. Rotate the tapes BD and CD about B and C respectively

93

until the 15-foot mark on CD coincides with the 12-foot mark on BD. Fix this point (D) with a peg. The angle ABD (CBD) is then a right-angle. (For construction of a right-angle in a small area BC may be made 3 feet, CD 5 feet and BD 4 feet.)

The perpendiculars can be extended as far as needed by the use of ranging poles, measured with a chain or tape and their length recorded. The perpendiculars are designated by reference to their distance from the end (origin) of the base line. Secondary perpendiculars (offsets) are then erected on the primary ones in the same way, at suitable intervals where they will 'pick up' (i.e. cross) the largest number of physical or vegetational boundaries. These offsets are recorded under the primary perpendiculars to which they belong and designated by reference to distance from the

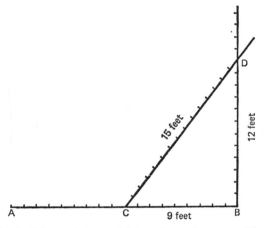

Fig. 9.1. Method of determining a right-angle on the ground (see text p. 93).

origin of the primary perpendiculars. The positions of boundaries or features crossed by the offsets, primary perpendiculars and base line are measured, recorded and a map constructed. The base line, perpendiculars and offsets are drawn to scale, the features of each line put in and joined up with those on adjacent lines. If the perpendiculars and offsets are sufficiently numerous and well located, the making of the map is straightforward, but if there is doubt about certain details, the map can be taken into the field (on a drawing board) and the boundaries completed by eye on the site.

Where plant communities are sharply defined and their boundaries simple, the 'chaining' method is quick, especially where data from the base line and primary perpendiculars only are sufficient for construction of the map. Where, however, many communities of complex distribution have to be mapped, a 'grid' method may be preferred. In this procedure, a complete system of squares covering the area to be mapped is erected on the base line. The side of the square is of any convenient size (say 100 feet, 50 feet, 20 feet, or in metric units), and the corners are marked by light rods or canes. Once the first squares have been constructed, any required number of squares can be quickly added by 'chaining' the intervals along two perpendiculars and 'ranging' the remaining corner rods in the two directions, i.e. parallel to and at right-angles to the base line. Each square can be mapped separately on squared paper, and should not be too large to be mapped by eye with reasonable accuracy. For larger squares, assistance from an observer on the side of the square with a measuring tape or rule, to give the recorder his position in the square, is helpful. By measurement relative to the sides of the squares, it is of course possible to locate any point precisely if required. With a little practice a suitable procedure is quickly reached.

At least two workers are needed to carry out the survey methods given above, but once the base line has been chosen, ranged and measured, a number of different parties can work on different parts of the area so that a considerable amount of survey can be quickly accomplished (e.g. in an afternoon). It is of course essential that independent parties should have precise and identical instructions as to the method of work, and the symbols to be used, so that the records may be uniform and complete.

The whole of the vegetation should, where possible, be carefully examined before the survey is started (e.g. during the previous afternoon), even when the species are known and a scheme of communities to be recognized decided. It is a mistake to set up more perpendiculars, offsets or squares than can be mapped in the available time, and it is not advisable to leave planted rods out overnight (except on private ground); if the area is not completed on one occasion the perpendiculars or squares can be quickly constructed if permanent pegs (flush or nearly flush with the ground) are left at intervals on the base line or elsewhere.

Much useful and appropriate survey can be quickly and accurately achieved by the use of a levelling telescope and levelling staff, especially if, as is often the situation, the boundaries of plant communities are strongly influenced by topography and level. Height above the water table is frequently important in relation to the distribution of a species, and often only small differences in ground-level are sufficient to affect the occurrence of plants.

For survey in this way it is convenient to lay out a base line, to set up the telescope at a convenient point on it, and to take readings by means of the levelling staff placed at desired distances along the base line which may be measured by a chain. Details of the vegetation at the sites are recorded as appropriate when they have been levelled. It is essential to adjust the telescope, which is set on a tripod, so that its optical axis is horizontal; this is accomplished by use of the spirit levels which form part of the telescope assembly. The levelling staff, normally in three sections extending to 15 feet, is held vertically and firmly with its base resting on the spot for which the measurement is required. A team of three members is desirable: one person holding the levelling staff, one reading the telescope and one acting as recorder. The distance of the levelling staff from the telescope need not necessarily be determined by chain, as it can usually be obtained directly from the telescope readings. Three horizontal hair lines, one above another, are visible on viewing through the telescope; the reading on the levelling staff given by the middle line indicates the height of the position of the staff, while from the difference in readings given by the upper and lower hair lines the horizontal distance between the levelling staff and telescope may be calculated. In this way the use of a chain can be dispensed with, a great advantage in some situations, especially where the ground is uneven and a substantial area is involved. By moving the position of the telescope tripod, the survey can be extended as far as required. Furthermore, sites away from the base line, and at any angle from it, can be easily levelled, and their vegetation recorded; their position is mapped from the horizontal distance and from the angle which the site makes with the base line (or other reference) which may also be readily determined by direct reading from the angular scale present on many levelling telescopes. If absolute heights of localities are required, referred to the datum of Ordnance Survey (the mean sea-level at Newlyn, Cornwall), the readings of level are continued to link up with the nearest Bench Mark, which can be found by inspection of the appropriate Ordnance Survey map. For many purposes, however, relative heights are all that are required, and these can be referred to an arbitrary datum. By means of the above procedures, a vegetation map linked with contours can be quickly obtained if the vegetation is recorded for the sites as their levels are being measured.

Yet other survey methods, involving for example a plane table or a dumpy level, can be employed, and such procedures are increasingly replacing the old-established 'chaining' methods.

It is rarely worth while to map a very large area in great detail, as time and effort are more profitably spent on detailed study, by means of quadrats, of small selected portions, or by transects across zoned vegetation (as described in Chapter 10). To this detailed work a more general map of the whole area may be added if required. For this the 25-inch (1:2500) Ordnance map may suffice as a basis, or a special survey on the lines described may be undertaken. Opportunities are thus available for exercise of the different interests and aptitudes of members in a class, the survey work being an excellent exercise in practical geography which can be combined with vegetation study. Such surveys and detailed recording are highly educational, evoking and training the powers of observation, encouraging accuracy, speed and ingenuity and giving the student close touch with the natural environment.

Chapter 10

Intensive Studies—Large-scale Charts of Vegetation

When we turn from extensive to intensive ecological work, we begin to come to grips with the detailed problems of vegetation, which cannot be solved by extensive work. Primary survey or the monographing on extensive lines of a given type of plant community over a wide area is largely a geographical study, concerned with the distribution and broad features of the larger units of vegetation. To understand the structure and development of individual communities, to learn how it is that particular species become dominant or abundant in some places and not in others, and to appreciate the mutual relations of species, how they compete, the advantages which they may confer on each other, their reactions to different factors of the habitat, and the limits of their tolerance of varying conditions, we must concentrate attention on particular problems and employ many different means for their solution. In this field of work, most particularly, flexibility, a lively imagination and ingenuity in devising methods of investigation are essential, as well as determination.

In dealing with primary survey one can describe fairly closely the necessary programme of work, but with intensive work it is impossible to do so. Because the problems are so varied and often differ according to the community studied, no fixed procedure can be specified. We shall, therefore, describe some of the methods that may be employed in studying firstly the vegetation itself and secondly the habitat. The methods chosen must depend on the particular vegetation to be examined and on the problems formulated in regard to it. Ecological techniques can sometimes be improved or profitably adapted to a particular problem in hand.

The methods described in this introductory book are mostly relatively simple, and do not involve elaborate training. If the need arises, more specialized methods can ultimately be adopted, but much of the behaviour of vegetation can be understood from studies involving simple procedures and equipment.

LARGE-SCALE CHARTS OF VEGETATION

An essential aid to the investigation and recording of vegetation data, though always to be used with discrimination and forethought, is the making of vegetation charts.

Vegetation charts represent the details of the vegetation of a small area on a large scale, and are constructed *de novo*. They are conveniently distinguished from *vegetation maps*, which are either constructed from pre-existing topographical maps by plotting the broad vegetational features on the topographical map as a basis, or by making a new map (Chapter 8). The vegetation chart is made to provide a characteristic sample of the detailed structure of a widely distributed typical community, to illustrate some striking distribution of vegetation in relation to habitat, or (and of great importance) as a definite aid to the solution of a definite problem.

Many striking distributions of plants are observed from time to time in the field—for instance, aggregations of one species around or among individuals of another, the growth of a species or a community only in positions exposed to or protected from the sun or the wind, the regular zonation of vegetation round a pond or lake, or again round a hillock or on the two sides of a ridge. Often such features can be sufficiently covered by descriptive notes alone or rough charts drawn by eye; an accurate chart to scale would not repay the time spent upon it. But in other instances the interest of the distribution and the accuracy with which it follows habitat conditions, which can often be separately determined and charted, merit an exact chart, and the close observation required for this purpose may reveal the existence of other factors not at first suspected, or may show that the supposed correlation with habitat, which seemed obvious, does not hold. The judgment required for a correct decision grows of course with experience in ecological work.

As a means to the investigation of a definite problem, charts are often indispensable. This applies especially to the study of succession or change in vegetation from year to year. The chart then becomes a record or datum with which future records can be compared; and only with their help can succession be accurately, and quantitatively, studied. We may often *infer* succession by comparing stands of vegetation and concluding that one represents a later phase into which another will in the course of time develop. When a number of obvious transition phases are available, such inferences have a high degree of probability, and most descriptions of successions are based on this kind of evidence. But at the best these inferences are not as satisfactory as direct observation on the same piece of ground. A succession which is directly traced has a certainty that cannot be questioned, the time involved is discovered, the exact details are followed, and the causes of succession are often automatically revealed. The drawback is, of course, that such direct study of succession has to be extended over several or many years; some successions occupy far more than the span of a lifetime. This difficulty may be partly met by choosing different stages of an inferred succession, as represented in different places, and studying each separately by means of a series of charts taken at intervals of time. In this way, if the supposed succession is a real one, the records may be continued until the last chart of one series corresponds with the first of another, and the whole may thus be pieced together.

THE GRID METHOD OF CHARTING

This is useful where large-scale mapping is required, where the boundaries of small, well-defined, uniform communities can be drawn, and where a record is needed of the approximate position of large isolated individual plants. Such charts are valuable, for example, in illustrating samples of vegetation in studies of succession, and where the distribution of small communities can be correlated with some physical factor, such as water content of the soil, or depth of water-level below the soil surface.

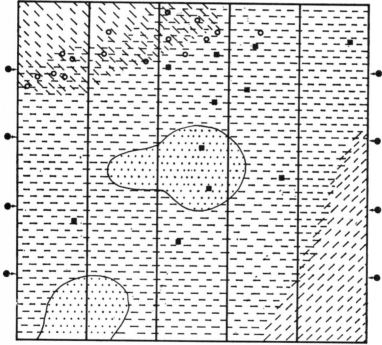

Fig. 10.1. Grid chart on a scale of 1 : 80 (actual charts should be larger than this). Three consociations and their transitions are represented by broken diagonal and horizontal lines and two examples of a society by dots. The positions of large isolated individuals of two species (■, ○) are shown, and of markers at intervals of 5 feet.

A square (or system of connected squares) is laid down, most conveniently on fairly level ground and often with sides of 5 metres, or of 25 feet. The corners of the squares are marked for future recharting by permanent pegs (creosoted hard-wood pegs driven in flush with the soil surface are desirable). If there is likely to be any difficulty in finding the pegs in the future, the exact position of at least one of them should be fixed by reference to two permanent objects in the neighbourhood to avoid waste of time in relocating the pegs on subsequent occasions.

When the two corner pegs of one side of the grid or square have been fixed at the approximate distance, a tape is run out at a right-angle from one of the pegs (see p. 93) and the square completed. Thin stakes can be stuck in the ground at convenient (often 1-metre or 5-foot) intervals along the sides of the square, and cross-tapes run parallel with two of the sides to divide the square into a series of strips. Alternatively, for more detailed mapping, two sets of tapes at right-angles can be used, the large square being divided into a number (often twenty-five) of smaller ones.

The grid can best be charted on squared paper. For areas of side 5 metres, a square decimetre of millimetre-squared paper is suitable, giving a scale of 1 : 50. For areas of side 25 feet, a square with 5-inch sides, ruled in tenths of an inch, gives a scale of 1 : 60. The boundaries of communities are easily and rapidly drawn in, and the names of communities can be indicated (e.g. by initial letters of the dominant genera) on the corresponding areas. These areas may be marked by interrupted lines in different directions. This has the advantage that transitional areas can be easily shown by the overlapping of two sets of lines (Fig. 10.1). The positions of any isolated large plants can be marked. To show all individual plants a much larger scale, not less than 1 : 10, is required. For indicating these individuals, the initial letter of the generic name is often a sufficient symbol; for different species of the same genus, the initial of the specific name can be added. The symbols used must always be clearly explained on the charts.

THE QUADRAT

The quadrat is a sample patch of vegetation of any desired size used for purposes of record. The term quadrat normally refers to a square area, readily enclosed within four tapes or laths or a frame for record purposes, the area often having a side length of 1 metre. However, forms of observational unit other than a square may be employed, such as a rectangle or circle, and a quadrat may now be defined as 'a sample area of given shape and size used for analysis within a plant community' (Hopkins 1954).

The simplest kind of record is merely to list the species enclosed within the quadrat, and so gain information about presence or absence. To this may be added the number of individuals of each species. As discussed in the next chapter, a large number of such quadrat records taken at random over a typical area of a plant community are necessary to determine its composition in quantitative terms.

The Quadrat Chart

The commonest size of a quadrat is 1 square metre, and for herbaceous vegetation this may conveniently be plotted on a scale of 1:10 (i.e. on a square decimetre of millimetre-graph paper). The boundaries of the quadrat may be made with tapes or laths, with skewers or pegs marking the corners. For successional studies the quadrat must be made permanent, by driving flush with the ground surface the four corner pegs of wood (treated with preservative) or metal. The position of the quadrat should be located by reference to at least two permanent and conspicuous objects in the neighbourhood, and an accurate record made of this location.

In charting close vegetation with many small plants, it is useful to lay a wooden metre scale parallel with, and 1 decimetre from, the bottom boundary lath or tape, so cutting off a strip 1 decimetre broad. A further shorter scale may then be laid 1 decimetre from the end of this strip, so that 1 square decimetre of vegetation is enclosed and can be charted by means of suitable symbols. Subsequently the next square decimetre along the strip is charted, and this process is continued until the strip is completed, when the metre scale is moved another decimetre up and so on until the whole quadrat is charted. By charting 1 square decimetre at a time in this way, maximum accuracy is obtained. When each square decimetre is charted, the 'joins' with adjacent squares may be checked with symbols already made on the chart. By starting at the bottom of the quadrat and working upwards, injury to the plants, from lying or kneeling on them before they are charted, is avoided. A permanent record of the vegetation of the quadrat may be required in addition to the chart itself and, because of possible damage to the plants, photography should be done before the charting is started. An alternative method of delimiting the sub-units of a quadrat is to use a metal frame with appropriate cross strands, or a wooden frame with strings. The advantage of permanent frames of this kind is that a quadrat can be very conveniently and quickly sub-divided simply by putting down the frame in an appropriate area of vegetation. Such frames are easily constructed in sizes up to about 1 square metre; for larger areas measuring tapes and posts are required.

For very close herbaceous vegetation, consisting of a great number of individual plants in a small area, a larger scale than 1:10 is needed. Usually a scale of 1:5 is satisfactory for such communities, but in extreme cases an even larger scale can be used and a quadrat smaller than 1 square metre may be desirable. In woody vegetation, however, such as scrub and forest, the quadrats must be bigger than 1 square metre. Quadrats of 5 and 10 metres square are often large enough, though in a mature forest of large trees even these sizes may not be sufficient and the grid method may be more suitable. It is seldom possible to include both trees and shrubs and also herbaceous ground vegetation in the same quadrat chart, because of the different scales involved. Large quadrats have less edge length in relation to their area than smaller ones, so that errors arising from deciding whether a certain plant is inside or outside the quadrat are smaller. On the other hand, large quadrats are more difficult to search than smaller ones and inconspicuous plants are in danger of being overlooked.

Tufted, tussock or cushion plants covering a considerable area should be outlined on the chart (see Fig. 10.2). The horizontal spread of the branches of plants casting considerable shade and of trailing branches may be indicated by suitable symbols. A distinct kind of vegetation, forming for example a definite stratum of the community, such as a moss or lichen

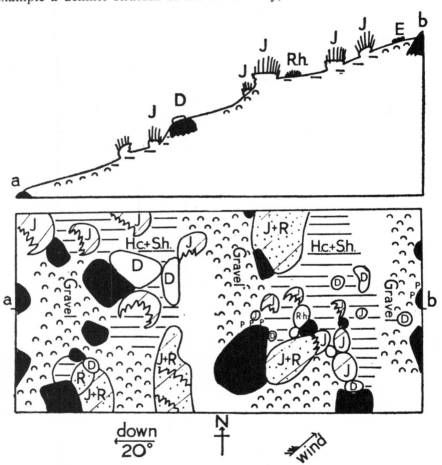

Fig. 10.2. Diagram of the structure of a terraced rush (*Juncus trifidus*) community in the Cairngorms, Scotland. The chart represents an area of 1 × 2 m. *Juncus trifidus* (J) occurs on the lips of each step shown in the profile diagram, and willow (*Salix herbacea*, S.h.) and a hepatic crust (H.c.) on the flatter part. Other plants present are the grass *Deschampsia flexuosa* (D), the crowberry *Empetrum hermaphroditum* (E), and the mosses *Rhacomitrium lanuginosum* (R), *R. heterostichum* (R.h.) and *Polytrichum piliferum* (P). Boulders are shown in black. (From M. Ingram, *J. Ecol.* 1958, courtesy of British Ecological Society.)

stratum, may also be indicated by a suitable symbol or shading. If such a layer contains a number of species and information is required about the detailed distribution of these components a separate chart for the particular layer can be constructed, although this is a rather laborious operation. Often, however, less detail is needed, and quadrats designed to show particular features, for instance the number and distribution of the individuals of certain species, may suffice and can be quickly recorded. Information on cover, or the proportion of ground occupied by particular species, is also often required and can be fairly quickly obtained from quadrat studies. Systems by which cover may be assessed are mentioned in Chapter 11.

Although most quadrats are square, other shapes of observational units are now used. A circular unit can be obtained by means of a metal ring which can be thrown at random within the area of a community (over the shoulder if it is desired to eliminate completely any possible unconscious choice of location). The species occurring within the ring are listed, and the operation repeated as many times as is necessary to determine the structure of the community (see Chapter 11). Large circular cbservational units are easy to operate in woodland, as only a central post and a string radius are required.

For charting purposes, however, a figure with straight sides rather than a circle is preferable and, while the square quadrat is the simplest and in some ways the most convenient, rectangles in the form of relatively narrow strips may, in heterogeneous vegetation, give more representative samples and more information about a community than square quadrats of the same size, although edge effect errors are higher (Clapham 1932, 1936).

Point Quadrats

In recent years, point quadrats have been much used, especially for studies on grassland and grazing. Groups of pins, often ten in a row about 5 centimetres apart, set in a frame through which the pins can slide freely and which keeps them vertical, are often employed. The apparatus can be easily made and used on uneven ground. Each pin point is essentially an extremely small quadrat, and presence or absence of a species is recorded by either a 'hit' or a 'miss' when the pins are let down, at random positions of the frame, on to the vegetation. Percentage cover can be rapidly assessed in this manner.

THE TRANSECT

This is the name given to a line or belt of vegetation selected for recording. The transect is particularly useful when the vegetation is *zoned*, i.e. when there are fairly regular successive bands representing different types of vegetation. Zoning may be related to a regular change in some factor of the habitat, such as decreasing water content with progressive distance from the edge of a lake, along a line perpendicular to the extension of the zones. It may also indicate a progressive invasion of plants into a seral community, without perceptible change in the habitat. The transect is made at right-angles to the zones, i.e. in the direction in which the habitat factors show maximal change or in the direction in which invasion is proceeding.

An advantage of the transect is that it shows a definite *range* of vegetation, and by recharting the transect at suitable intervals of time any progressive change in the vegetation along the line of the transect can be established.

The Line Transect

This is the simplest form of transect made by running a measuring tape or chain along a desired line covering the distance over which records are required. The position of the plants touching the tape are recorded, usually by the use of suitable symbols.

The scale of the transect chart is chosen in accordance with the size of the individual plants and the closeness of the vegetation. For a record of the individual trees and shrubs in a wood, for example, a scale of 1:50 or 1:100 may be used, but for herbaceous vegetation 1:10 is suitable, except where the plants are very small, e.g. crowded annuals springing from seed on fallow soil or in a salt marsh, and here a larger scale may be required.

The Belt Transect

This is a strip of vegetation of uniform width, which can be marked out by two parallel tapes or chains, and the vegetation between them recorded (Fig. 10.3). The width of the transect (e.g. 10 cm, 25 cm, 1 m, 6 in, 1 ft) is selected as appropriate according to the type of vegetation. Transects passing through woodlands need to be wider (perhaps 5 m) than those in close uniform herbaceous vegetation of small plants, where a decimetre may be quite wide enough.

Records can be made in the same way as for quadrats. In fact the transect can be regarded as a series of quadrats placed end to end, and the belt transect combines the advantages of the line transect and the quadrat. While the belt transect is used chiefly in the study of detailed changes of vegetation along its length, it is wide enough to show the distribution of plants in two dimensions. As with the quadrat, the transect can be marked permanently and observations can be repeated over an interval of time to investigate changes in vegetation.

PROFILE CHARTS

The Stratum Transect

This is a profile of the vegetation, drawn to scale, and is primarily intended

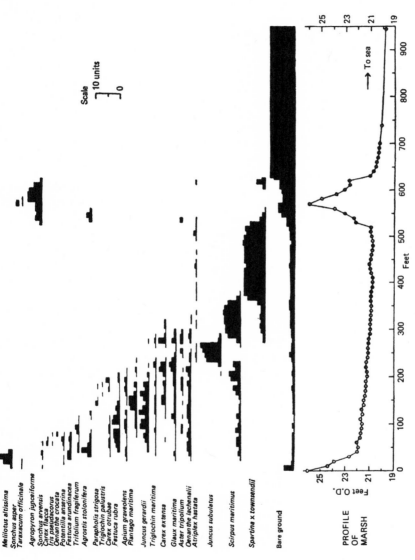

Fig. 10.3. A belt transect across a salt marsh and a newly-building small dune seaward of the marsh, Berrow, North Somerset. The vegetation was assessed by relative bulk (see p. 110) in successive areas 10 ft (3·05 m) long and 3 ft (0·91 m) wide. The distribution of most of the plants is shown by the histograms, but species of rare occurrence are omitted. (From A. J. Willis and E. W. Davies, *Watsonia*, 1960, courtesy of the Botanical Society of the British Isles.)

to show the relative heights of the plant shoots. It is based on the line transect, and complements the belt transect, adding the vertical dimension to the plants.

A measuring tape or chain may be used to mark the line of the transect, and the heights of the plants touching the tape or chain measured and recorded. With tall plants, it is convenient to run one or more horizontal strings at definite heights above the tape for ease of measurement. A pictorial semi-diagrammatic representation of the vegetation can be made, which is especially useful where transitions between communities are involved. The vertical vegetation profile indicates the shoots of all the plants on the transect, drawn as informatively as possible. If the diagram also includes such features as water-level, topography and soil type, ecological factors which may have a strong influence on the vegetation may be suggested. This is a most valuable and instructive exercise.

For some vegetation it is useful to base profile diagrams on a belt rather than a line transect. In woodland a line transect would include very few tree trunks, but a profile based on a belt of, for example, 5 m wide, gives considerable information (see Fig. 3.1).

The Bisect

This is a stratum transect chart in which the root systems are included as well as the shoots (Fig. 10.4). Ideally a trench should be carefully dug by the side of the transect line, the root systems of the plants isolated, and the vertical and lateral spread plotted to scale. Great care is needed to avoid breaking the finer roots. Usually it is practical to study the detailed distribution of the roots of only a few individuals of different species, because of the difficulties involved. Nevertheless root structure and distribution are of major importance in understanding plant ecology, and it is often instructive to excavate and to try to isolate the root systems of particular plants. The extent of root branching is often very different in different types of soils and in soil of different water content, as shown, for example, by the extensive and instructive studies of Weaver (1919) in America (Fig. 10.5). The size of the root relative to the shoot and the extent of the vertical penetration of the root vary considerably from species to species, as seen, for example, in plants of chalk downland and sand dunes described by Salisbury (1952). Whereas the root systems of many dune plants are very deep, those of some of the ground flora of woodlands may be relatively shallow, roots of different species often occupying different layers of the soil (Fig. 10.6). Exploitation of different soil layers by roots is well exemplified by the ground flora of certain oak woodlands in which bracken (*Pteridium aquilinum*), bluebell (*Endymion non-scriptus*) and soft-grass (*Holcus mollis*) occur together. Excavation shows that in shallow soils the soft-grass roots near to the surface, the bracken rhizomes and roots are deeper, and still

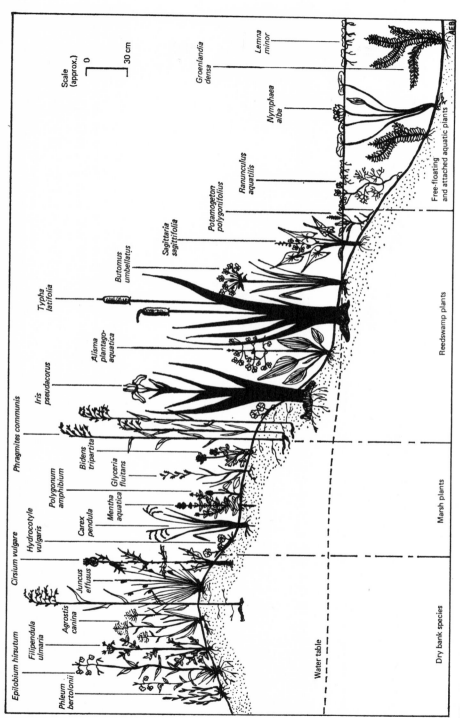

Fig. 10.4. Bisect from a dry bank through a marshy area to a pond, illustrating the different forms of the plants growing in different conditions. The vegetation is shown much simplified and semi-diagrammatically.

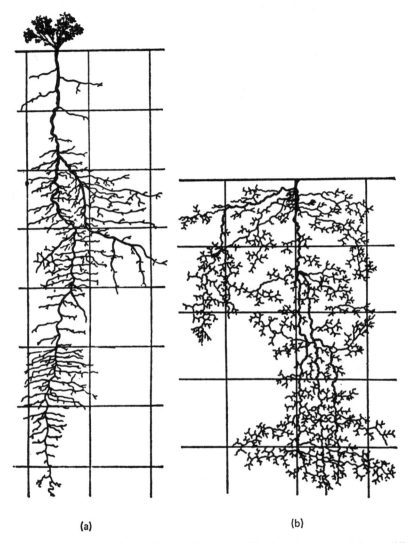

(a)   (b)

Fig. 10.5. Root systems of two plants of a spurge (*Euphorbia montana*) from different habitats in North America, showing a wide difference of development. (From J. E. Weaver, 1919, Publications of the Carnegie Institution of Washington, 286.)

(a) Plant from the very compact dry soil of the Great Plains. The root penetrates more than 7 feet (the squares are of side 1 foot), with poor development of absorbing rootlets except in the 3rd and 4th foot, where a fissure in the soil held water seeping from the surface, and in the 6th foot where the soil was moist and easily permeable.

(b) Plant from 'gravel slide' on the Rocky Mountains. Here the soil is very coarse, loose and easily permeable. The rootlet system begins near the surface and extends only to the 5th foot, but the lateral spread is considerable (often more extensive in the uppermost foot than shown).

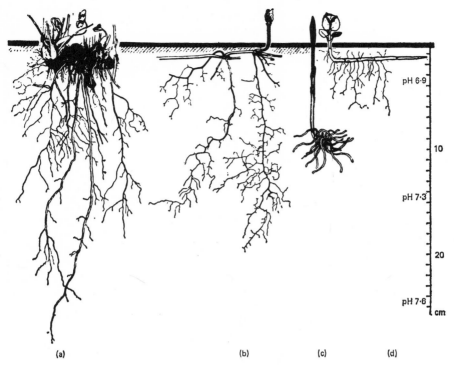

pH 6·9

10

pH 7·3

20

pH 7·6

cm

(a)   (b)   (c)   (d)

Fig. 10.6. Root systems of plants of dry beechwood: (a) wood sanicle (*Sanicula euro-paea*); (b) dog's mercury (*Mercurialis perennis*); (c) white helleborine (*Cephalanthera damasonium*); (d) enchanter's nightshade (*Circaea lutetiana*). The soil reaction changes with depth and is alkaline below the surface layers. (From E. J. Salisbury, 1952, *Downs and Dunes*, courtesy G. Bell & Sons.)

deeper are the bluebell bulbs, the roots of the components of this plant society not directly competing with one another.

THE USE OF VEGETATION CHARTS

It is important always to bear in mind the object in making vegetation charts and records and to choose appropriate procedures for the particular study in hand. Detailed information about the structure of plant communities is essential for understanding their ecology. Communities are not static entities but are frequently changing, both by cyclical processes and by succession, as discussed in Chapter 4. Only by intensive and repeated study of vegetation can we learn how it came into existence, how it maintains itself and how it may progress. Such study by means, for example, of quadrats and transects is of great educational value, as it leads attention to the details of the vegetation; yet the graphic record should not be regarded as an end in itself, but the mind should always be kept open to the problems raised and efforts made to solve them.

THE PERFORMANCE OF PLANTS AND THEIR RELATIVE ABUNDANCE

In addition to the detailed charting of vegetation, it is often desirable to obtain some measure of the yield of plants in defined areas and the relative contribution of the various species present to the total bulk of the vegetation. Such information is required especially if the effect of some factor on the growth of plants is being studied, either in natural conditions or experimentally. It may, for example, be desired to investigate the effect of light intensity at the edge of a wood on the yield and floristic composition of the vegetation, or of the effects on vegetation of increasing the levels of mineral nutrients in the soil. In the former case it would be necessary to sample the vegetation from just inside the wood to an open area just outside it, and in the latter to make observations before nutrient addition and at appropriate time intervals afterwards.

The charting procedures already described could be used to provide information about the distribution of the plants, but would not, for example, give much indication of the greater amount of plant material in the ground flora in sample areas outside the wood as compared with inside it, or the increased growth of plants often resultant on increases in nutrients.

A simple way of obtaining an assessment of yield is to cut all the plant material at ground level (with large scissors or shears) from a defined area, such as a quadrat or sub-unit of a quadrat, and weigh the cut vegetation. This can be done in the field, with a fair degree of accuracy, by putting the harvested fresh vegetation in polythene bags and weighing on a spring balance (screening from wind is desirable). The dry weight of the vegetation can subsequently be obtained, if the material is dried in an oven and re-weighed; dry weight provides a better index of yield than fresh weight as clearly the latter varies with conditions of soil moisture and with the incidence of rainfall. It is a fairly simple matter to obtain the fresh weight of the total vegetation of a sample area, but it is also possible, though requiring care and patience, to obtain the yields of all the individual species, if they are separated into different bags for weighing. In this way an accurate measure of the contribution of each species to the total vegetation in terms of weight is gained.

A quicker estimate of the relative contributions of each component to the total bulk of the vegetation can be made by eye, without destroying the vegetation by harvesting it. In this subjective method the relative bulk of each component is judged visually in a sample area, and an appropriate allowance can be made for any bare ground on an area basis (Willis, Folkes, Hope-Simpson and Yemm 1959). Tiny plants occurring sparsely in quadrats may be recorded as 'trace' amounts, and the major species are assessed according to their size and number. Comparisons of the subjective estimates with those obtained from cutting and weighing show good agreement (Willis 1963) and, when experience has been gained, the subjective method is quick and informative. Procedures for determining the abundance of species by objective methods are given in Chapter 11.

Other measures of the performance of plants may often provide valuable information. The average height of vegetation is usually fairly easily determined, as well as features of individual plants such as numbers and heights of inflorescences, and, for some plants, numbers and lengths of runners. Vigour is not so readily assessed, but it is sometimes possible to develop a suitable arbitrary scale for a particular species based on the general appearance of the plants.

Chapter 11

Quantitative Methods of Vegetation Analysis

While broad generalized descriptions of vegetation may sometimes be all that is required, for close comparisons of stands of vegetation and, for example, for the detection of small changes in the composition of vegetation, exact and rigorous procedures are needed.

The use of abundance classes in which a 'frequency' symbol is assigned to each species recorded in a community based on a subjective judgment is described in Chapter 8 (p. 89). Although this procedure is suitable for quick reconnaissance surveys, it is not recommended for intensive studies on small areas. The interpretations which different observers place on the different terms and on the visual estimation of abundance in the field are liable to differ; different persons, following a particular recording system, may reach different results, and, as already noted, even a single trained observer may obtain different records for the same area on different days (Hope-Simpson 1940; Smith 1944). Consequently it is important that objective methods should be used which give repeatable results.

A valuable first exercise is to consider possible methods in the field, by going to a nearby plant community; a lawn or playing field is ideal for this purpose. Two difficulties in the objective and quantitative determination of the abundance of the various components of the vegetation will soon become apparent. In the first place the whole community is too large for a complete study in any reasonable time; and secondly, many species spread vegetatively in such a way that it is not possible to recognize individual plants and to assess the number present.

SAMPLES

To record the whole community is virtually impossible, but small samples (quadrats) may be selected for detailed analysis. Such samples must, however, be representative of the community as a whole. This would not be so if they were all placed near one corner of the community or localized in groups here and there; neither should they be distributed in a regular manner. Each sample should be taken at *random*.

Samples from any population enable two measures relating to each character considered to be estimated: the mean and the scatter from the mean, called its variance. The *mean* is the arithmetic average. The *variance* is a measure of the extent to which the individual values, from which the mean is calculated, differ from the mean. Where the individual values differ only little from the mean the variance is low, but where the individual values differ considerably from the mean the variance is high. For example, the individual values 3, 4, 5, 6, 7 and 1, 3, 5, 7, 9 have the same mean, 5, but the second set has the larger variance. The mean and the variance are important characteristics of each population, but unless the samples are distributed at random the variance cannot be easily determined. Furthermore, if the variance is unknown it is impossible to compare populations statistically.

Several methods are available to locate samples at random. One is to use the distances along two adjacent sides of a rectangular area as ordinates and co-ordinates and place the position of successive samples by means of random numbers which can be obtained from tables. This rather slow procedure can be speeded up by the use of pacing instead of actual measurements.

Sample areas or quadrats from any population will differ from one another because of the inherent variation in the population, but estimates of the mean and variance will increase in accuracy with increasing numbers of samples. In other words, the *sampling error* of these estimates can be reduced by recording more samples or quadrats. The sampling error in fact decreases in relation to the square root of the number of samples and is independent of the proportion of the total population sampled. Ecologically, this means that it is generally better to observe a large number of small quadrats than a smaller number of larger ones, but clearly a decision on quadrat size also depends on the nature of the vegetation (see p. 101), the object of the study, and the accuracy required.

Normally records of at least 25 quadrats are desirable and for precise studies 50 or 100 are necessary. Point quadrats (see Chapter 10) have a very high variation because of their very small size and consequently a large number, about ten times that for other quadrats, should be recorded.

CRITERIA OF ABUNDANCE

Essentially, there are four measures which may be used in assessing the abundance of the various species present in quadrats: *density*, *cover*, *frequency* and *yield*. These measures relate to distinct features of the vegetation and have their own particular advantages and disadvantages.

Density
This is the number of individuals per unit area, and ideally is the best feature to assess. The number of individuals of the species, or of each species independently, present in each quadrat is recorded and the average number per quadrat determined. This average can easily be converted into

a mean number for any area (m², ha, acre, etc.) so that results obtained from quadrats of different area can be readily compared.

One difficulty over the use of density is that all individuals of the same species count as equal; a sprouted acorn and a mature oak tree both rate as one individual. For herbs, however, size differences are generally relatively unimportant, and for trees it is possible to work in terms of basal area rather than density—a method often employed by foresters.

A more serious difficulty is, as already noted, that for many species individual plants cannot be recognized. This is so for the large number of herbs which spread vegetatively by means of such organs as rhizomes and stolons. For these plants, shoots, suckers, tillers, and even leaves have sometimes been taken as 'individuals' for the purpose of density counts.

Cover

This is the proportion, or percentage, of ground occupied by a species, or more precisely by the vertical projection of its aerial parts. In any closed plant community, because of the overlapping amongst species and the different strata of vegetation, the combined percentage cover for all species exceeds 100; only in some open communities is the value less than 100 per cent.

Cover is not an entirely objective measure and is normally determined by estimating the approximate cover of the various species as they occur in each quadrat. There is considerable danger of the underestimation of small- and fine-leaved species (Hope-Simpson 1940). Nevertheless cover is a useful measure of species abundance. As it is expressed as a percentage it is independent of the size of the quadrat, so results obtained from different quadrat sizes are readily compared. In woodlands it may also be possible to record tree cover, or canopy, by looking vertically upwards above the quadrats.

Very fine point quadrats, whose cross-sectional area is negligible, give a less subjective determination of cover. The percentage of such points touching each species is equal to the cover of the species. An even more accurate method is to use a cross-wire apparatus which gives a true point.

Often a system of cover classes, which simplifies field recording and subsequent calculation without much loss of information, is used. One such system (Trepp 1950) is as follows:

Class	Observed range of cover (%)	Mean cover (%)
+	<1·0	0·1
1	1·0– <10	5·0
2	10– <25	17·5
3	25– <50	37·5
4	50– <75	62·5
5	75–100	87·5

In the field each species is recorded in each quadrat in terms of its cover class; the mean percentage cover can be calculated later for each species by use of the values shown above.

A somewhat similar and more widely used system is the Domin scale. This has eleven classes as follows:

Isolated, cover small	×
Scarce, cover small	1
Very scattered, cover small	2
Scattered, cover small	3
Abundant, cover about 5%	4
Abundant, cover about 20%	5
Cover 25–33%	6
Cover 33–50%	7
Cover 50–75%	8
Cover >75%	9
Cover about 100%	10

Despite the large number of classes, results obtained by the Domin scale may be less objective than those obtained by the first scale, as it is not based entirely on cover; the lower values depend partly on frequency (see below). Moreover, values are usually estimated once only for each species (for the whole community without the use of samples). Nevertheless, the Domin scale is practical to apply, especially in simple communities, and different observers generally reach the same results by its use.

A scale similar to the Domin scale involving both cover and frequency is that developed by Braun-Blanquet (see Braun-Blanquet 1932) and used by many ecologists on the continent of the Zurich–Montpellier school. The scale is as follows:

Sparsely or very sparsely present, cover very small	×
Plentiful, but of small cover value	1
Very numerous or covering at least 5% of the area	2
Any number of individuals covering $\frac{1}{4}$–$\frac{1}{2}$ of the area	3
Any number of individuals covering $\frac{1}{2}$–$\frac{3}{4}$ of the area	4
Covering more than $\frac{3}{4}$ of the area	5

The Braun-Blanquet and Domin scales avoid the difficulty of assessing the cover of numerous small individuals and of deciding what constitutes an individual in the many plants which spread by vegetative means.

Frequency

In its strict sense, frequency is the chance of finding a species in any one quadrat; the observed frequency is the proportion, or percentage, of quadrats in which a species is present. It is the most widely used method of indicating the abundance of species.

The field procedure is to observe a number of quadrats and record how many of them contain the species. In practice all the species present in each quadrat are usually recorded and the frequency of each species is then calculated. This is a very simple and quick operation. All that is necessary is to determine (identify) the species and decide whether to use presence as rooted presence (the plant rooting in the quadrat) or shoot presence (the plant's shoot occurring vertically above the quadrat).

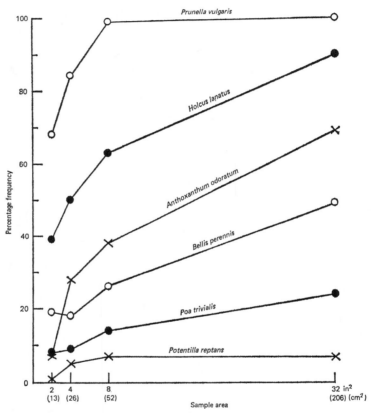

Fig. 11.1. The relationship between frequency and sample area for a number of grassland species. (From data of G. E. Blackman, 1935, courtesy of *Annals of Botany*.)

Although frequency is often recorded in the field, being easy and quick to assess, there are disadvantages in its use. For example, frequency depends on quadrat size and shape. The results obtained from one quadrat cannot be directly compared with those from another of different area or shape, for although frequency increases with sample area the relationship is not a simple one. Fig. 11.1 shows the results for several species from grassland communities sampled with quadrats of different areas. Frequency increases with increase in sample area but the relationship varies from species to species. The relationship between frequency and sample area depends not only on the density of plants but also on the degree of clumping or aggregation of individuals. If individual plants were distributed at random there would be a predictable relation between frequency and sample area, but in the field plants are normally aggregated to a greater or lesser extent. Consequently density cannot be determined from observed frequency values, as frequency measures not only the abundance of a species but also its pattern of clumping.

Another disadvantage in the use of frequency is that the values are of low precision; a species must have two very different frequency values (on two areas measured by the same type of quadrat) for there to be a statistically significant difference between them. The following presents the range of reliability of selected frequency values obtained from 50 samples (Greig-Smith 1964):

Observed frequency	0	10	20	30	40	50
Range of reliability (95% confidence limits)	0–4	5–17	14–28	22–36	33–45	46–50

Thus for a species with an observed frequency of 20 in 50 quadrats, in 19 out of 20 cases the 'true' frequency will lie between 14 and 28. This overlaps with the frequency reliability range of from 22 to 36 for an observed frequency of 30. Consequently, observed frequencies of 20 and 30 out of 50 quadrats are not significantly different, and the true frequencies of observed frequencies of even 20 and 40 from 50 samples may not be very different. An increased number of samples improves the precision but not substantially: for 100 quadrats the observed frequencies of 40 and 60 (the equivalents of 20 and 30 for 50 quadrats) are only barely significantly different.

Yield

Because it is laborious to determine, yield (see Chapter 10) has not been very widely used in ecological studies as a measure of abundance. Nevertheless it is a very reliable index, especially if the amount of plant material is determined by its dry weight, and this measure, although involving the destruction of the vegetation, is a valuable one to obtain at the end of experimental studies. Fresh weight is, as noted earlier, subject to variation dependent on rainfall and the water content of the soil.

It will be clear from the foregoing that measurements and comparisons of the abundance of plant species are far from simple. Like many biological data, field observations have to be treated statistically in order to yield valid results. The method of such treatment cannot be given here, but reference may be made to elementary books on statistics for biologists, such as those by Campbell (1967) and by Clarke (1969), to the works by Greig-Smith (1964) and by Kershaw (1964) which treat quantitative plant ecology in some detail, and to the book by Phillips (1959) on methods in plant ecology.

Attention has so far been given to abundance, but there are other important quantitative aspects of plant distribution, and some of these are considered below.

VEGETATION ORGANIZATION

Pattern
Observation of the distribution of individual plants in the field, such as dog's mercury (*Mercurialis perennis*) in a wood, and the red poppy (*Papaver rhoeas*) in an arable field, soon reveals that different species are distributed in contrasted ways, some plants being clumped together, like the dog's mercury, but others, like the poppy, not aggregated in this way. Individual plants may in fact be seen to be distributed in one of three manners, shown in Fig. 11.2, usually referred to as random, regular and aggregated. In a *random* distribution each individual occurs strictly according to chance and independently of all other individuals present. In *regular* distribution individuals are arranged in some uniform order, such as that in which trees are often planted. An *aggregated* distribution results when individuals occur in groups, being clustered or clumped together.

Various procedures are available to determine whether plants are distributed randomly (and much statistical theory depends on drawing a random sample from a randomly distributed population). Although the basis of the methods is a little involved, the procedures are relatively simple to use. The principles of the two main types of methods are outlined here, and details may be found in the works by Greig-Smith (1964) and Kershaw (1964).

One procedure involves the use of random quadrats. By reference to those shown in Fig. 11.3 it will be seen that all the quadrats on the regularly distributed population contain one individual. However, on the aggregated population of the same density, sampled by quadrats of the same size and shape (Fig. 11.3c), most quadrats contain no individuals and one contains many individuals. The quadrats on the random population (Fig. 11.3b) contain numbers intermediate between those of the other two types. It can be calculated, for a given density and quadrat area, how many quadrats would contain each number (0, 1, 2, 3, . . .) of individuals if the population were distributed at random. The calculated and observed values are then compared. By further reference to the diagrams (Fig. 11.3) it will be seen that, in comparison with the random population, the aggregated population has more empty quadrats and the regular population fewer.

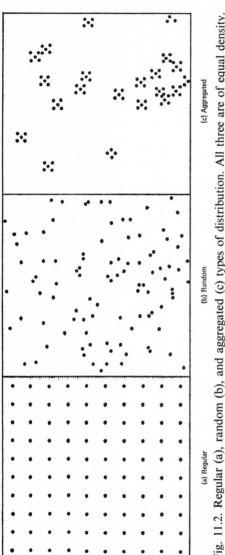

(a) Regular   (b) Random   (c) Aggregated

Fig. 11.2. Regular (a), random (b), and aggregated (c) types of distribution. All three are of equal density.

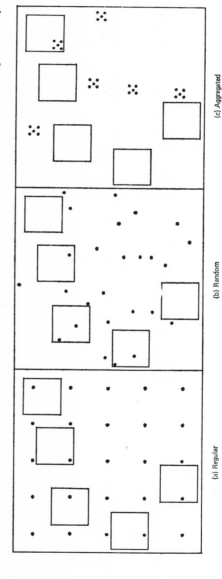

(a) Regular   (b) Random   (c) Aggregated

Fig. 11.3. Part of the populations shown in Fig. 11.2 sampled by five random quadrats. The mean number of individual plants per quadrat is seen to be the same for all three types of distribution (a–c), but the number of empty quadrats differs (see p. 118 for further details).

The principle of the other group of methods concerns the distances between individual plants and their nearest neighbours and between random points and their nearest plants. In a completely regular population the maximum distance a random point can occur from a plant is less than the plant-to-plant distance (Fig. 11.4a). On the other hand, in an aggregated population the plant-to-plant distances average considerably less than the point-to-plant distances, for the random points will often fall in the gaps between the groups of plants (Fig. 11.4c). In the random population the plant-to-plant and the point-to-plant distances are equal (Fig. 11.4b).

Analysis of a large number of species from many types of vegetation shows that the vast majority of plant populations are aggregated. Random populations are occasionally found where a species is invading a homogeneous bare area. Regular populations, which may be expected in conditions where there is intense competition for a certain factor (e.g. water in deserts), occur extremely rarely, largely because of unevenness (heterogeneity) in the environment.

The reasons for plant aggregation are, of course, of great interest to the ecologist. Differences in the environment are often important, and even when they are only slight they may determine the germination, establishment and growth of plants. Many plants exhibit clumping as a result of their method of growth and mode of vegetative reproduction, e.g. thyme (*Thymus drucei*) and wood anemone (*Anemone nemorosa*). Even in species which normally reproduce and spread by seeds, the tendency is for most of the seeds to fall near to the parent plants. Because of these different reasons the individuals of most plant species are aggregated or exhibit what has been called *pattern*.

The study of pattern has received much attention recently (Greig-Smith 1964), particularly with a view to discover the scale of pattern and to determine its biological causes. The techniques involved largely depend on sampling by quadrats of different sizes and comparing the number of individuals which they each contain.

Association Between Species
The pattern exhibited by individual species is only one aspect of the complexity of organization found within plant communities. Most species are known to be aggregated, and the question arises as to how the patterns exhibited by the various species in the same community relate to each other, and of what significance they are in the community as a whole. The investigation of the association between species has been mainly carried out by comparing the joint occurrences or abundances of pairs of species in quadrats.

For a pair of species of known frequencies the number of quadrats in which they both would occur, if they were distributed independently of

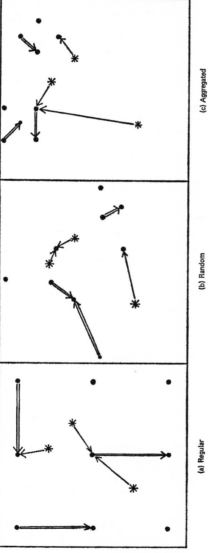

(a) Regular   (b) Random   (c) Aggregated

Fig. 11.4. Part of the populations shown in Fig. 11.2 sampled by three random points (shown as stars). Note the differences between the mean distances from a random plant to its nearest plant (double arrow) and from a random point to its nearest plant (single arrow) for the three types of distribution (see text, p. 120, for further details).

each other, can be easily calculated. The *expected* joint frequency is given by:

$$\frac{\text{frequency of species A} \times \text{frequency of species B}}{\text{number of quadrats recorded}} .$$

The expected value is compared with that observed by a simple statistical method (using a 2×2 contingency table and testing with χ^2) to determine if the two species are associated to any significant extent. This procedure is repeated for all pairs of species (rare ones are often omitted) and the significant associations determined. Associations, of course, may be positive or negative according to whether the two species occur together more or less often than expected by chance. From such results, groups of positively associated species are determined.

The associations between species in a north Pennine bog are illustrated (Fig. 11.5). In this study seven groups (A to G) were determined and the ecological role of each was assessed (Hopkins 1957). The main bog surface was covered by plants of group A; other groups occurred as patches scattered amongst this group. For example, the plants of group E grew in wet hollows whilst the *Sphagnum* spp. (groups C and D) were units of the active growing bog surface.

The detection of such mosaics helps to show the structure of communities and may suggest that the phases delimited are cyclically related in time (see Chapter 4).

Association-analysis and Ordination

In a heterogeneous community it is particularly useful to be able to classify the quadrats into groups; it is then often possible, by field observations on these groups, to suggest (and subsequently test by experiment) ecological factors which may be responsible for their distribution.

By the technique of association, quadrats from a single community, or a series of related communities, can be classified into groups (the method is laborious and normally carried out by computer). The degree of association is calculated for all species pairs and the species which has the greatest amount of association is selected. The quadrats are then divided into two groups: those containing this species and those without it. The whole procedure is then repeated for each of these two groups separately and again repeated in all further groups until either the associations become insignificant or sufficient groups are obtained. This method is called *association-analysis* (Williams and Lambert 1959) and results in arranging the quadrats into groups.

The value of association-analysis is well exemplified (Fig. 11.6) from observations made at Shatterford in the New Forest (Williams and Lambert 1960). A total of 504 quadrats containing 29 species was analysed and gave 15 significant groups (only the 6 most significant are shown here).

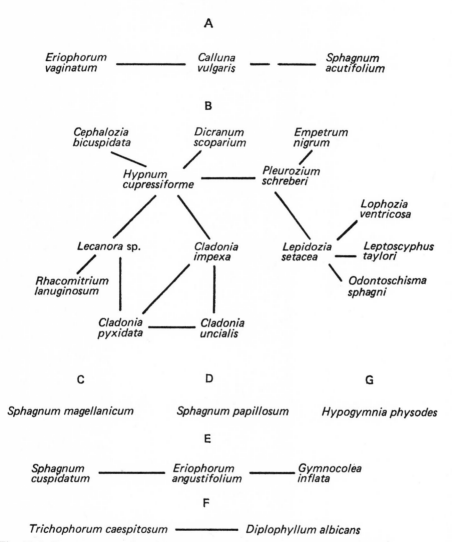

Fig. 11.5. Seven groups of positively associated species (labelled A to G) from a north Pennine bog community. (From B. Hopkins, *J. Ecol.* 1957, courtesy of British Ecological Society.)

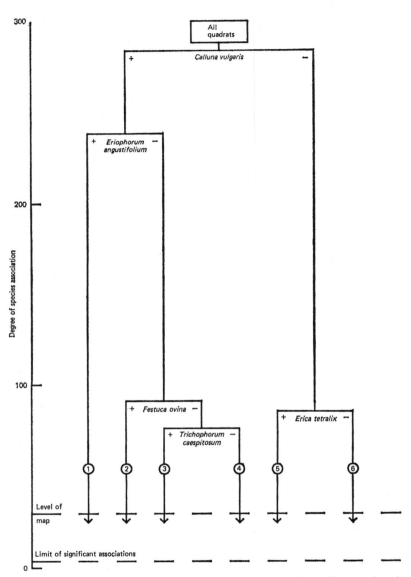

Fig. 11.6. Association-analysis of vegetation at Shatterford, New Forest, showing the six most significant groups obtained. (Simplified from W. T. Williams and J. M. Lambert, *J. Ecol.*, 1960, courtesy of British Ecological Society.)

The quadrats were arranged in a rectangular area, almost 100 metres long, down a slope with a fall of about 4 metres. The soils of this area, the summer limit of surface water and the main vegetation types are shown, together with the distribution of the six groups, in Fig. 11.7. The most significant division (on ling, *Calluna vulgaris*) separated the bog (without *C. vulgaris*) from the heath communities (with *C. vulgaris*), a boundary which closely follows that of the summer limit of surface water. The

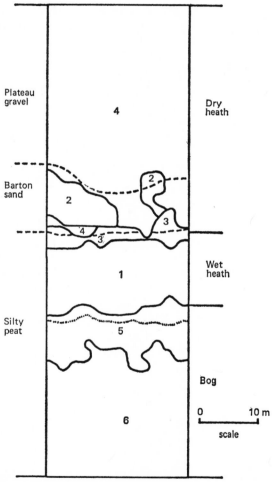

Fig. 11.7. Map of Shatterford, New Forest, showing the distribution of the six most significant groups obtained by association-analysis (numbered 1 to 6 (see Fig. 11.6) and separated by continuous lines), the soil types (indicated by broken lines, names on left), the position (dotted line) of the summer limit of the surface water, and the main vegetation types (names on right). (Simplified from W. T. Williams and J. M. Lambert, *J. Ecol.* 1960, courtesy of British Ecological Society.)

second division (on cottongrass, *Eriophorum angustifolium*) delimited wet heath on silty peat (with *E. angustifolium*) from dry heath on mineral soils (without *E. angustifolium*). Thus the first two divisions of the association-analysis separated the vegetation into its three major types of dry heath, wet heath and bog. The other divisions are at a much lower level of significance (see Fig. 11.6). Of the three dry heath groups, group 2 (with sheep's fescue, *Festuca ovina*) mainly occurs on Barton sand, group 4 on plateau gravel, and group 3 (with deer sedge, *Trichophorum caespitosum*) along the edge of the wet heath. The bog groups were divided on cross-leaved heath, *Erica tetralix*; those with *E. tetralix* adjoined the wet heath.

In contrast to this classificatory method of association-analysis is the technique of *ordination* which is valuable in indicating the main ecological factors responsible for the distribution of stands of species.

In ordination studies, field observations of the abundance of species in a series of stands from related communities are analysed (usually by computer), all pairs of stands being compared. Ordination diagrams show the plots arranged so that the distances between them are proportional to their degree of dissimilarity, and inspection of the diagrams often indicates the ecological factors which are of overriding importance.

TRANSITIONS BETWEEN COMMUNITIES

The basic techniques for the analysis of transitions between plant communities are the same as those already described for the analysis of a single community. In the analysis of a community the samples studied are distributed at random; but in the analysis of transitions they are taken in a row (*transect*) perpendicular to the community boundaries (p. 104).

The transect is subdivided into a number of plots whose length should normally be not less than the transect width. These plots may either be treated as individual samples or subsampled by random quadrats. In both cases the analysis may be carried out by any of the methods previously described.

CHANGES IN TIME

The methods of analysis so far discussed all record the vegetation at a particular point in time. Vegetation, however, is, as already stressed, not static but dynamic and it is often desirable to record changes which occur in time. Information may be required about changes at different seasons of the year and also over a period of years.

Such studies may be made by sampling the same plots repeatedly by use of the same sampling methods. A difficulty which may arise here is that there may be more variation within the samples taken at a single time than between samples taken at different times. One solution to this problem is to set up *permanent sample plots*. These must be clearly marked by corner posts which can be found on subsequent visits. Permanent plots may be either charted or analysed quantitatively, for example, by means of point quadrats (p. 103). This is a suitable approach to adopt in following any cyclic change or succession along a zonation such as a hydrosere.

If general changes in time are being investigated in a single community by this method, there is a danger that too much reliance may be placed on sampling certain small areas only. Even if these were originally chosen at random they were still a sample of the whole, and any difference they showed from the mean of the whole community may persist throughout the time in which they continue to be recorded. Thus there is considerable danger that permanent sample plots may not be typical of the community which they represent. This problem can be overcome by recording plots on two occasions only. On the second occasion a new series of plots is established and recorded. These plots too are abandoned after their second recording, when a further set is established, and so on. In this scheme, changes on the same spot are recorded but each plot is recorded only twice and so no one area is over-recorded.

Chapter 12

Individual Species and the Parts They Play in Forming Communities

It was said in Chapter 3 that the study of a plant community always and necessarily drives us back to the individual species, and we begin to realize that there is still much to learn about many of them. Often one of the first things of which we realize our incomplete knowledge is the growth and behaviour of plants at different stages of their life history (especially at the important phase of seedling establishment) and the reproductive capacity of the different species in the conditions under which they grow in nature. The fundamental importance of intensive and detailed autecological study cannot be overestimated.

MAINTENANCE AND DISPERSAL OF SPECIES

Take, for instance, the herbaceous vegetation of a wood. What do we know of the means by which the different species maintain themselves from year to year, or spread from one place to another? In any given species, does it regularly ripen seed and in what quantity? Different species vary enormously in the amount of ripe seed which they produce and in the viability of their ripe seeds. It has been shown (Salisbury 1942) that on the whole a species produces more than enough seed to ensure survival and to have a chance of increasing its range when opportunity offers. Plants growing in shady places ripen fewer seeds, but these are usually larger and have a greater stock of reserve food, so that they are more likely to produce new plants that can establish themselves successfully. Generally the largest seed productions are characteristic of species of habitats which are available for colonization only intermittently, such as woodland clearings and the exposed mud of shallow lakes and ponds (Salisbury 1942).

Does the seed of a particular species fall and germinate on the spot, and do the seedlings establish themselves and grow into plants, which flower in their turn? Is seed spread to new areas, and does it thus increase the distribution of the species? All these questions require specific answers based on exact observation, for the process of reproduction by seed may be interrupted at any point in the series of processes.

The maintenance of perennial species—and nearly all woodland plants are perennials—is effected by the persistence of the original plants from year to year, or by the outgrowth of rooting offsets, or by some other means of vegetative reproduction. In addition to this, new plants may be produced, so far as there is room, by seed falling and germinating among the parent plants. On the other hand, dispersal from one place to another must take place by the actual carrying away of some part of the parent plant, and in the great majority of cases it is the fruits or seeds which are so carried.

Many seeds or fruits appear to have definite aids to dispersal: for instance, plumes or wings which, by offering a greater surface and therefore increased resistance to the air, cause the seed or fruit to fall more slowly to the ground, and thus enable it to be carried further by the wind before it alights; hooks or some sticky substance which may attach the fruit or seed to some passing animal, by which it is carried; or a fleshy envelope which is eaten by a bird or mammal, the seeds being afterwards voided. Species are dispersed in all these ways, but the 'dispersal mechanisms' are by no means always effective, and the seeds of plants possessing them are not always carried by the agents (wind or animals) that seem appropriate. We must also note that a vast number of species have no special means of dispersal.

The questions which we have to answer in studying the spread of vegetation are not concerned primarily with general categories of 'dispersal mechanisms', but with the actual ways in which particular species are dispersed in particular places; and this is a problem—by no means always an easy one—which can be solved only by direct observation in the field. Careful field observations often show that dispersal may be brought about in several ways; for example, in the bluebell (*Endymion non-scriptus*) the primary dispersal mechanism is the ejection of the seeds from the dry capsule, but the activities of various animals, including rabbits and badgers, may occasionally serve to give somewhat greater dispersal (Knight 1964).

It is clear that seeds or fruits which are carried to a distance from the parent plant do not all germinate, and if they do germinate the seedlings may not succeed in establishing themselves. Many factors, including seed dormancy, temperature regime and soil conditions, notably moisture, influence germination and especially its timing (see, for example, Newman 1963); furthermore, soil conditions, both physical and chemical, strongly affect seedling establishment (Rorison 1967). The vast majority of viable seeds that do not reach suitable soil in which to germinate are permanently lost. Sometimes large unmistakable seeds or fruits, such for instance as acorns or beechnuts, may be seen lying on the ground in places where they are very unlikely ever to germinate, or, if they do germinate (owing, for instance, to continuous heavy rain), to survive. Young beech seedlings have been found in chalk grassland at some distance from the nearest parent trees, but on a dry soil so extremely shallow above solid chalk that the primary root soon dried up and establishment was impossible. And acorns in great masses are found in every good acorn year on ground where they have no chance to establish themselves.

The death rate of seedlings, like that of all young organisms in nature, is enormous. Both beech and oak seedlings, after good beechnut and acorn years, are found in immense numbers on the floors and on the edges of woods. But the vast majority disappear in the course of a few weeks or months. Systematic effort can throw much light on the causes of this (Watt 1919, 1923) and similar failures of seedlings to establish, as, for example, those of ash (Wardle 1959, see p. 134).

The sort of situation often seen is the colonization of a suitable new habitat by a species at some distance from the nearest parent plants, while unsuitable habitats at an equal distance remain uncolonized. The most obvious hypothesis to explain this common occurrence is that seeds from the parent plants are spread everywhere, at least to that distance, but that all die except those which reach the suitable habitats. This may very often, but not invariably, be true. There may be some dispersal factor which takes seed only or mainly to the particular new habitat. For example, a new tree plantation on arable or grassland may receive seed from the woodland plants of neighbouring woods carried by woodland animals or birds, or on the clothes or boots of woodmen, beaters or sportsmen, as shown long ago by Woodruffe-Peacock (1918); the seeds of plants with many different types of dispersal mechanism may be carried on footwear (Clifford 1956). Sometimes what appear to be equally suitable new habitats remain uncolonized. On the other hand, the dense initial colonization of the common ling (*Calluna vulgaris*) in suitable habitats indicates widespread dispersal of seeds by the wind, and that they germinate and establish seedlings only in suitable spots. Here the dense colonization may be explained by the high seed production (which may be more than 100,000 seeds per plant per year) and the high seed viability (Gimingham 1960) but for many species information on seed production and seedling establishment is incomplete.

A knowledge of the means of maintenance of a species in a place where it is already established can often be obtained by direct observation at different times of the year. Do the already-established plants persist from year to year? Do any of them die and, if so, why? Does the species regularly or occasionally produce ripe seed, and do these seeds fall close to the parent plants, germinate, and produce new plants which successfully establish themselves? Sometimes all this information can be obtained by simple observation. A permanent quadrat, charted in succeeding years, will give accurate quantitative information as to the appearance of new plants from seed and the disappearance of the original ones.

One method of studying dispersal is by observing the appearance of new species where they did not exist before. In the first place, one has to be quite certain that the species was not present all the time, though so hidden or inconspicuous as to be missed. After a coppice is cut, for instance, many species which have been dormant or nearly so in the deep shade of the fully grown coppice, represented mainly or entirely by rhizomes or other under-ground organs, burst into vigorous growth and flowering. Disturbance or ploughing of an old soil may bring to the surface, so that they germinate, seeds which have been deeply buried and dormant for many years. New plantations on old arable land or old grassland, and new habitats of all kinds, are particularly favourable for establishing the fact of migration from a distance.

The actual agents of carriage are sometimes hard to determine. Much, however, can be learned about wind carriage by observing the transport of winged and plumed seeds and fruits during gales of exceptional force. Occasional gales may be very important in distribution. The distance to which winged and also small light seeds and fruits can be carried by the wind is often considerable; aerial eddies may take small seeds and spores to great heights where they can be carried by wind for long distances before falling. Records show that the dust-like seeds of orchids may, exception-ally, travel as much as hundreds of miles, and fungal spores can reach heights of more than 10,000 feet (3000 m). Wind-borne seeds and spores are known to fall on ships substantial distances from land, and plants coloniz-ing remote islands are often those with wind-borne seeds (Ridley 1930). Close observation of the habits and food plants of animals, both wild animals, especially birds, and also cattle, sheep, and horses, will often give a lead that may be successfully followed. Seed transport by birds, both externally such as in the mud adhering to the feet of water birds, and internally, serves to spread a considerable range of plants, as is clear from the weed species found associated with gull colonies (Gillham 1970).

Man himself as a carrier of seeds is by no means to be neglected. The importance of man in the past in the distribution of plants is shown by the number of weeds and ruderal (wayside) species first recorded in Britain in the Neolithic and Bronze Ages (Godwin 1956). Man probably introduced many such species from the Continent, and also helped to create conditions favourable for their growth in this country. In modern times, man has promoted the dispersal of a large number of plants as impurities of crop seeds and many alien species have been brought to Britain with imported material, notably wool (Salisbury 1964).

Short distance dispersal can be studied by observing the extension of a species from year to year. Favourable cases for investigation have to be looked for carefully, for many species in many habitats are almost station-ary. When a favourable instance has been discovered, the rate of advance can be measured by permanent quadrats or transects of suitable size on the edge of the area, and the means of dispersal can usually be established in the course of the observations of these. The advance of a perennial species is often effected almost entirely by the growth of rhizomes or runners.

COMPETITION AND THE ESTABLISHMENT OF NEW CONDITIONS

In a community of perennial plants the number of fresh seedlings that can establish themselves among their parents is limited by the available space. This does not necessarily mean, of course, that the number of plants can increase until the shoots are in lateral contact. The root systems of plants growing spaced out from one another may occupy the whole intervening soil and so use all the available water in the soil, so that though there is physically room for more *on the surface*, new individuals cannot establish themselves, though the seeds may germinate. In this way we may get an *apparently open* community in stable equilibrium with its habitat. With a greater water supply, more individuals, of the same and of other species, are able to come in, until eventually we get a *closed* community in which the shoots are in lateral contact. The competition is then for space and light. If tall plants form part of the community, there will be room below them for lower-growing plants, and thus we get the beginning of stratification. But these lower-growing plants must be able to grow successfully with illumination less than full light from the open sky, for some of the light is cut off by the taller plants. When species of several different heights come in, the layering is increased, and in a temperate forest, as earlier noted, there are commonly four strata—trees, shrubs, herbs and mosses; however, where the tree canopy is very dense, so much light is cut off that shrubs and sometimes even shade herbs are unable to grow and the vegetation consists of one or two strata only. As the number and variety of species and the bulk of vegetation increase, more humus accumulates in the soil, so altering its character, and rendering it suitable for other kinds of plants.

In this way we get increasing differentiation, increasing complexity of the community, somewhat parallel with the differentiation and increase in complexity of an advanced animal or human community, where also there exist different categories of members playing very different parts in the life of the community as a whole, but all dependent on the total food supply available.

Eventually, however, a limit is reached, determined partly by the number of species in the neighbourhood and able to reach the community, and partly by the structure and economy of the community itself. For these exclude species unable to fit into that structure and economy when it is once well established on account of the limitations of light and water supply, the constitution of the soil as modified by the existing plants, and other factors. The later stages of development of a climax community are often marked by a decrease in the number of species, since many of those that flourish in the middle stages of development, where the conditions are intermediate and very varied, are unable to subsist under the ultimate more extreme and more uniform conditions, for instance the deep shade of a wood in close canopy with its damp atmosphere and richly humic soil.

The processes which have gone to the making of such a complex community, given the power of arrival of the species which compose it and the general nature of the habitat, are two: competition and the establishment of conditions by some of the species which enable certain other species to exist. In general, the members of different layers do not compete because their shoots and very often their roots occupy different strata.

It is desirable to study these processes and to trace out in detail exactly how they lead to the building up of the community (Watt 1924, 1925). It is well to start by trying to gain a broad understanding of the structure and economy of the climax, the adult community, just as in studying a type of organism it is well to start from the adult form. But we cannot fully understand the significance of all the features of an adult except in the light of a knowledge of its development. Similarly the understanding of a complex community requires a knowledge of the way in which the vegetation has developed, and the factors affecting it during the course of succession.

The methods appropriate in a study of this kind are very varied. Charting by the various procedures already described is essential for an accurate and detailed knowledge of the structure of the various stages, and the information obtained in this way often leads straight to a closer understanding of the processes involved in succession—competition and the establishment of new conditions. But observation of the facts of succession is constantly making us ask what precisely, in quantitative terms, are the modifications in conditions which lead to the disappearance of one species and the appearance of another, for it must be remembered that every change of conditions affecting the life of plants can be expressed ultimately in terms of chemistry and physics. In other words, we want to measure the change in habitat factors and to determine which are effective in changing the community. Habitat factors are considered in succeeding chapters, but for their complete elucidation long continued work, requiring glasshouse and laboratory studies, is necessary. However, much can be learned by a thorough study of succession, aided by quite simple and straightforward observations on habitat, and wherever possible by field experiment.

The possibilities of field experiment to determine crucial points are almost unlimited, and can frequently yield much important information. Thus where water supply may be crucial, small patches of ground can be artificially drained or watered (Farrow 1917*b*; Jeffreys 1917), and so valuable qualitative information gained. Field conditions may sometimes be profitably simulated by controlled large-scale experimentation, as, for example, in studies on the purple moor grass (*Molinia caerulea*), shown to grow better in conditions of flowing than of stagnant ground water (Webster 1962).

Root competition is sometimes of crucial importance. The shallow roots of the trees in a beechwood, for instance, may so drain the surface layers of soil that few herbaceous plants, or none, may be able to establish themselves. This point can be tested by making a quadrat between the trees and cutting off their surface roots by digging a narrow trench along its boundaries. This may enable herbaceous plants or undershrubs to establish themselves within the area of the quadrat (Watt 1931). A filled-in trench in a forest floor may support many tree seedlings and other species absent from adjoining areas, demonstrating the marked effect of root competition.

The effective value of different light intensities in excluding or admitting species can sometimes be tested by sowing seeds in variously shaded parts of a wood and noting their germination and subsequent growth. Some plants can produce seedlings, but cannot permanently establish themselves under certain degrees of illumination. For example, seedlings of ash (*Fraxinus excelsior*) appear in abundance in woods in most years, but it has been shown (Wardle 1959) that they fail in dense dog's mercury (*Mercurialis perennis*) where the light intensity is below their compensation point (the intensity at which the energy income of the plant just balances its energy outlay). In shade other plants can vigorously develop their vegetative organs, but cannot flower or ripen seed. If the plants flourish and set seed perfectly well, some factor other than light, such as, for instance, difficulty of dispersal, must be responsible for their absence where there is room and the habitat is otherwise suitable. Many woodland plants can flourish quite well in the open provided that they have a sufficient water supply and the atmosphere does not become too dry during their growing season.

The effects of animals can often be determined by preventing their access in various ways (Farrow 1916; Watt 1919, 1923). Large animals such as sheep can be precluded from an area by means of a fence, and so grazing by these large animals is eliminated completely or else permitted at a controlled intensity. In one such study on hill grazings in north-east Scotland, Gimingham (1949) showed that in many ungrazed sites heather (*Erica cinerea*) dominates ling (*Calluna vulgaris*) but that moderate grazing shifts the balance in favour of ling. The success of the ling under grazing is attributable to its production of spreading branches, whereas the upright branches developed by the heather when the main shoot is bitten off are readily vulnerable to further grazing.

The effects of rabbits can be studied by construction of a rabbit-proof enclosure. The occurrence of myxomatosis in Britain in 1954–5, severely reducing the rabbit population, gave a unique opportunity for the assessment of the effects of rabbits on vegetation on a countrywide scale, further information becoming available with the eventual return of the rabbits in some places (Thomas 1960, 1963). From such studies the great effects of rabbits on the growth and composition of vegetation is evident; in the absence of rabbits palatable grasses and woody plants are found to increase.

In this chapter we have done no more than touch upon some of the main problems raised in the intensive study of vegetation, but it is hoped that enough has been said to indicate the enormous and varied field involved, and some of the ways in which the study may best be approached.

Chapter 13

Photography of Vegetation

It need hardly be said that photographs of plants or of vegetation are of no scientific value unless there is some definite purpose which they fulfil with at least some measure of success. A large proportion of the photographs taken, and even a number of those published, are of little or no value from any point of view, i.e. they are neither instructive nor beautiful. Many subjects are photographed under conditions which offer no prospect of success, and even when the conditions are relatively favourable the film is often carelessly exposed. Students of vegetation who use photography should concentrate on the production of a few really good photographs, each with a carefully considered and definite aim.

There are two main scientific reasons for taking a photograph of vegetation: firstly the desire to make a picture of a characteristic sample of some definite type, secondly to make a record for the purpose of comparison with other records, for instance with a future photograph of the same spot or with a quadrat chart.

Under the first head we have the snapshots of the student engaged in reconnaissance or primary survey. Good snaps of characteristic landscapes showing the kind of country and including one or more typical plant communities are interesting and useful, and the more successful photographs may be used to illustrate a published account. It is impossible to lay down hard and fast rules for the taking of these. Landscapes showing good contrast in bright diffused light are likely to give the best photographs. Though some subjects, especially distant ones, photograph well in bright sunlight, this is generally to be avoided where the vegetation is at all close to the camera, because the excessive contrast and heavy shadows will probably obscure the form of the plants. It is essential to focus carefully and, if the camera is hand-held, to use a fast enough shutter speed to avoid the risk of camera shake. When a tripod is carried, or the camera can be rested on a support, a longer exposure may be used and a greater depth of field obtained by stopping down the lens. If there are plants or significant details of vegetation in the foreground it is best to make sure that these are sharp and to leave the background to take care of itself. Wind is often a problem in photographing vegetation. As far as possible, photographs of vegetation should be taken in calm weather, and a sufficiently short exposure used to arrest any movement of leaves and branches that there may be.

136

Photographs of excellent quality can be taken of vegetation with 35 mm and roll-film cameras if these are carefully used. When time is available for taking more leisurely photographs, a large technical camera taking sheet film, used on a sturdy tripod, will yield the finest results. A camera of this kind has the advantage that, by using the swing back, objects in a landscape from the foreground to the far distance can be rendered sharply on the negative without excessive stopping down. Whatever camera is used, the importance of good composition and satisfactory contrast may be emphasized. A good pictorial effect is by no means negligible from the scientific standpoint. A pleasing and effective picture impresses the features of the subject on the mind much more strongly than an ugly or poor one. The best weather and time of day to secure favourable lighting should be carefully chosen. Vegetation generally shows up best in a photograph if it is lighted laterally, and not from above or from directly behind the camera.

For portraits of individual plants, it is worth taking some pains to select an individual that composes well within the format of the photograph, that falls within the narrow depth of field available when working close up, and that stands out well from its background. Diffused lighting generally gives the best results. Patience is needed to wait until the plant is perfectly still before making the exposure. For plants of a dimly lit woodland floor, flash or a mixture of flash and daylight illumination may be desirable.

A good medium-speed fine-grain panchromatic emulsion is suitable for photographing vegetation. It should be borne in mind that the vegetation usually shows some heavy shadows and, if detail in these shadows is to be recorded, then sufficient exposure must be given. In such a situation, the maxim is 'expose for the shadows and let the highlights look after themselves', though care should also be taken to avoid over-exposure, especially of 35 mm films.

Filters can often be employed to advantage. A light-green or yellow filter, requiring $\times 2$ or $\times 3$ increase of exposure, will lighten the tone rendering of the green colours of vegetation, and will also help to bring out the clouds in a landscape photograph. A polarizing filter, correctly orientated, can be useful in photographing submerged vegetation and other features which may be obscured by reflections from a water surface.

For general purposes an MQ-borax developer is satisfactory and can be used for almost all black-and-white emulsions. For 35 mm films, carefully exposed in good quality equipment, 'compensating' or 'high definition' developers can be helpful in retaining shadow detail without producing excessive highlight density, and they render sharp detail with a pleasing crispness. An MQ-borax developer diluted 1 : 3 with water gives something of the same effect. Over-development of photographs of vegetation should be avoided.

COLOUR PHOTOGRAPHY

Colour photographs are in many ways easier to take than photographs in black and white, and although, for reasons of cost, colour photographs can rarely be used to illustrate published ecological work, colour transparencies are invaluable for recording features of vegetation. The main pitfalls stem from the fact that the form of the final transparency is determined at the moment the exposure is made: there is little or no opportunity to rectify errors later. Correct exposure is important with colour reversal (transparency) films; an error of more than 1 stop (× 2) from the correct exposure is unlikely to give an acceptable result. Careful attention to composition at the time of exposure is particularly important in the case of colour transparencies. The relatively large and accurate viewfinder provided by the focusing screen of a reflex camera is a great help in composing well-balanced photographs which make full use of the picture space.

Particularly effective colour photographs of vegetation can often be taken during the autumn and winter months, when the tints taken on by different species are often much more strikingly differentiated than the relatively slight variations of green during the growing season.

STEREOSCOPIC PHOTOGRAPHY

Stereoscopic photography is very valuable for the recognition of individual plants in a community, though the necessity of viewing these photographs through a pair of lenses detracts considerably from their value as a means of illustration. The appearance of relief seen in stereoscopic photographs is a great help in recognizing individual plants which would be scarcely separable in an ordinary photograph. Stereo-photographs are best taken with a stereoscopic camera, but they may also be obtained by making two successive exposures with an ordinary camera which has its optical axis moved laterally by 2–3 inches (5–8 cm) between the two exposures. By the latter method exaggerated relief may be obtained, or fairly distant objects such as the trees in a wood may be made to stand out better than they would in photographs taken with an ordinary stereoscopic camera, the amount of relief depending on the distance through which the optical axis of the lens is moved.

QUADRAT RECORDS AND ORIENTATED PHOTOGRAPHS

In photographs which are simply scientific records it is rarely possible to consider pictorial effect, and even a poor photograph may be better than none at all, though naturally the best possible under the circumstances should be the aim. In this category come photographs of quadrats, transects or areas which are being mapped. The quadrat boundary laths, transect tapes, etc., should always be included. They give a certain definiteness to the photograph which is both useful and effective, and they may also help to give an indication of scale. In general photographs it is often desirable and sometimes important to include a well-known object (such as a coin, pen-knife, trowel or ruler) in the picture which can serve as a guide to scale.

In photographing a quadrat the camera is best placed a little outside the bottom boundary, tilted forward so as just to include the length of the front lath, focused on the middle of the quadrat and stopped right down. The quadrat will appear as a trapezium, but the vegetation will be less foreshortened than if the camera is horizontal. When the whole of the vegetation included in the quadrat is very low, forming a carpet with no plants rising much above the general level (e.g. a turf or moss community), it is a good plan also to photograph the quadrat, or part of it, with the camera pointing vertically downwards. This can be managed with the help of a ball and socket 'universal joint', screwing to the top of the tripod.

Sometimes it is desirable to take photographs at an appreciable height, which is greater than that possible by holding the camera by hand, or from support on a tripod. A vertical photograph may, for example, be needed of a large quadrat or the vegetation of an area of more than some 10 sq ft (1 m²). For this, a camera mounted on a pole constructed of a number of sections may be used, as described by Jones (1955), the shutter being operated by a long string; photographs above a water surface can also be obtained by means of this device. An alternative is to use an arrangement of two legs forming an inverted V, which can be hauled up by means of a rope, the camera being sited in the angle of the V.

Permanent quadrats, especially of communities with well-marked aspects such as woodland ground vegetation and meadow land, should be photographed in each aspect, i.e. several times in the season. At the least all permanent quadrats laid down for the purpose of studying succession must be photographed once a year—if possible when the vegetation is at its maximum luxuriance, and always at the same date, or at least within a few days of the original date.

Photographic records of vegetation which are intended for comparison with others of the same spot taken at earlier or later periods should be photographed with the same camera in *precisely* the same spot at *precisely* the same height and pointing in *precisely* the same direction. Even a slight deviation in any of these respects will render *exact* comparison impossible. The best way to secure this result is to drive a permanent peg into the ground exactly under the middle of the camera, and a taller stake so that it comes exactly in the middle of the picture, and can be focused upon. If a note is made of the height of the lens above the ground, succeeding photographs which are strictly comparable can then be taken.

AERIAL PHOTOGRAPHY

Photography of vegetation from the air has in recent years produced remarkable results. Not only does it provide a rapid and accurate method of recording the distribution of different types of vegetation, but species of dominant plants can often be identified in photographs taken from a height of several thousand feet.

Low altitude aerial photographs at a scale of 1:1000 (or larger) permit plant identification from leaf characters, but the appearance of tree-crowns is often distinctive on a much smaller scale (1:25,000). Not only can coniferous and deciduous trees be readily distinguished, but often individual types of tree can be determined; for example, holly (*Ilex aquifolium*) trees show up as small dark canopies, whereas oak (*Quercus robur*) is evident as somewhat irregular lighter canopies with the clustered foliage of the major branches visible.

Aerial photographs, both vertical and oblique, are of particular value to the ecologist for several reasons. They often show clearly the effect of contrasting soil conditions on vegetation in respect of density, growth, type and colour of plants. They may also show features which are not easily seen on the ground, such as the sites of former boundaries and buildings, or evidence of ploughing. Further, they are of great value in primary survey, especially in inaccessible areas, and can provide an up-to-date, permanent, detailed record of transient features of the vegetation, shown in their true relationship, of sites which are not readily amenable to ground survey, e.g. dense forests, cliffs, salt marshes (Plate 12), floating reed beds. For rapid mapping of major vegetational zones, aerial photography is unequalled, yielding information not easily available through time-consuming ground survey. Nevertheless, it is very desirable, wherever possible, to correlate air photographs with observations made of features on the ground.

Seasonal and also long-term changes of vegetation, as well as the colonization of bare areas such as mud flats, and the rate of spread of plants, as, for example, of the cord-grass (*Spartina* × *townsendii*), can be recorded by repetitive aerial photography. A further special use of aerial photography is the detection of disease in trees. Diseased leaves, invaded by fungi, reflect much less of the infra-red part of the spectrum than healthy leaves, and are easily detected by the use of infra-red film or by 'false colour' film.

Vertical air photographs are now available for essentially the whole of Great Britain, often spanning several decades, so that not only can the present features of vegetation, soil type, land-use and topography be made out but also the photographs give an indication of the immediate past vegetational history of the area.

Part IV
THE HABITAT

Chapter 14

Climatic and Physiographic Factors

In modern ecological work the term *habitat* may be taken to mean 'the sum of the effective conditions under which the plant or the community lives'. Originally it meant the *place* in which it lives,[1] but while the word is still commonly, and of course quite legitimately, used in this sense, it has now become a scientific term applied to all the conditions affecting the plant incidental to the place in which it lives. Consequently we have to distinguish the general habitat of a community from the particular habitat of an individual plant belonging to it, for it is at once obvious that the conditions under which an oak tree lives are different from those of the moss growing upon its bark, though they have some points in common, for instance the general climate of the locality.

Every species and every community has a certain *range* of habitat, which may be wide or narrow. Some species are distributed over a large portion of the globe under a considerable variety of climates, while others are confined to a very restricted set of conditions which may be realized only within a small area. It by no means follows that such a species can live only within the area to which it is actually confined. It may not have completed its natural migrations, and may still be in course of extending its range. All species tend constantly to increase their range, and it has been shown that the areas covered by species vary on the whole with their age, i.e. with the time during which they have been in existence.

But this tendency is by no means always realized. A species may be prevented by various kinds of barriers—such as oceans, mountain ranges and deserts, or closed plant communities which it cannot enter—from spreading outwards from the area to which it is restricted, though if it is

---

[1] From the Latin *habitat*, 'it lives in' or 'inhabits', e.g. *Primula habitat in silvis*, the primrose lives in woods: hence the habitat of the primrose is woodland.

141

transported across these barriers to a suitable spot it will establish and propagate itself. This is well seen when European plants are transported to temperate North America or to New Zealand, or when plants from the Old World tropics are carried to the tropics of the New World, and vice versa. Plants which successfully colonize new areas in this way frequently have rapid growth rates and produce many, easily dispersed, often wind-carried seeds in a short time, i.e. are 'weeds'. If, for example, a stretch of land in virgin country is ploughed, weeds will appear, establishing and flourishing in the open soils, though they could not cross the space separating the newly ploughed land from their nearest habitat without the aid of man. Thus we may distinguish *actual* from *potential* habitats.

Theoretically, of course, we should be able to analyse the different factors of the habitat into the ultimate physico-chemical forces acting upon the plant, but knowledge in the field of physiological ecology is far from complete, despite substantial advances in recent years. Much, however, is now known about the physical features of the environment: for example, the variation of humidity and its effects on plants, the variation of water content of the soil and its ecological significance, and the range of temperature, both seasonally and diurnally and in different parts of the habitat. Knowledge of the chemical features of the environment such as the composition of the soil, its hydrogen ion concentration, and the levels of essential mineral nutrients and their effects on the occurrence and performance of plants, is also substantial.

It is important to recognize that vegetation cannot be understood adequately without knowledge of the conditions of the habitat. It is not sufficient to know merely that a plant is to be found in a certain habitat and that where it occurs certain other plants are also likely to be found; we need to know also why this plant and its associates are found in this habitat and excluded from others. A detailed and exact study of the structure and distribution of vegetation should therefore be complemented by a full consideration of the physical and chemical features of the environment, for only by such considerations can the distribution of plants be satisfactorily interpreted. The study of certain features of the habitat is complex, involving elaborate apparatus, but much information of value can be gained by the use of simple equipment and methods such as those outlined below.

For general purposes we may conveniently group habitat factors (ecological factors) into *climatic, physiographic, edaphic* and *biotic*, though these are not always sharply separated, and there are very many interactions of these different factors.

Climatic factors include the general features of regional climate and season, light, temperature, rainfall, humidity of the air and winds; but they may also vary to an important extent locally (local climate) and even in extremely restricted areas (microclimate).

Physiographic factors are those determined by the general nature of the geological strata, by topographical features, such as altitude, slope, aspect and exposure, and by geodynamic processes such as erosion, silting and the blowing of sand.

Edaphic factors are those dependent on the soil as such, involving its physical and chemical constitution and such features as water content and aeration.

Biotic factors are those due to living organisms, either animals or plants.

It will easily be seen that the factors assigned to different classes act and may react upon one another. Thus the climatic and physiographic factors influence one another, and both affect the edaphic and biotic, so that some factors are largely dependent on the others. The soil water content, an important edaphic factor, is primarily influenced by climate, particularly the amount of precipitation, and the rate of evaporation, the latter being strongly dependent on atmospheric humidity, wind and insolation (incidence of radiant energy from the sun). Soil water content is, however, also influenced by physiography, particularly the slope and nature of the soil surface, by other edaphic factors such as the physical structure of the soil, and by biotic factors, especially the nature and the density of the vegetation which not only affects the amount of rainfall reaching the soil but is constantly drawing on the soil water through transpiration. In general, factors classed as climatic have a dominating influence upon all the others, and temperature has been described as a 'master factor' in the distribution of vegetation.

Rainfall, the lithological nature of the strata ('rocks' in the wide geological sense), and their varying altitude, together determine the size and course of the rivers and streams, and thus the conformation, slope and exposure of the land. Altitude, slope, aspect and exposure determine what is called 'local climate', affecting the temperature, rainfall, atmospheric humidity and insolation to which a given stand of vegetation is exposed, and therefore to a large extent the particular species of plants that form it. Marked *local climates* exist on summit ridges, on steep northern or southern slopes, in deep ravines which are relatively cool or damp, and in shallow basins into which cold air drains and in which it lies during calm weather. On a smaller scale *microclimates* are created by the physical presence of rocks, or even a large stone, which may protect the immediately neighbouring plants from insolation or from wind. Vegetation creates its own microclimate by intercepting radiant energy and by reducing the strength of the wind, so that in a stand of vegetation plants of low stature experience less extreme climatic conditions than those of taller growth. This is as true in grassland vegetation as in the more obvious case of woodland vegetation, and has an important bearing on the measurement of climatic factors.

Physiography and climate also interact with the rock strata to determine in large measure the nature of the soils which are formed, and thus the

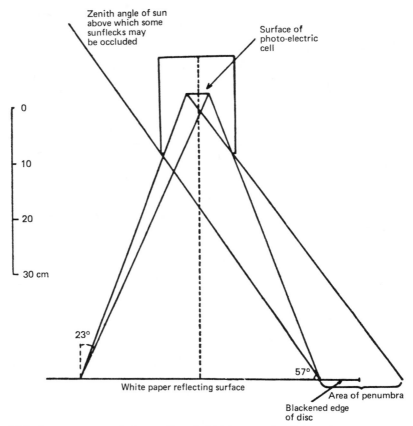

Fig. 14.3. Arrangement of photocell and shield to receive light reflected from a horizontal white surface of known area. (After G. C. Evans, *J. Ecol.* 1956, courtesy of British Ecological Society.)

for example, for use in tree canopies (Jackson and Slater 1967). Calibration can be carried out by use of a light meter or against a Kipp solarimeter although the latter measures total radiation whereas the photometer measures radiation essentially between the wavelengths of 0·40 μm and 0·64 μm. Caution is therefore needed in interpretating results from this instrument.

The spectral composition of light may be investigated by using inexpensive colour filters, made of gelatin film, which transmit a sharply defined waveband. They are best used in conjunction with a radiometer, since the relatively narrow sensitivity of other instruments imposes a considerable restriction.

A useful measurement of radiation conditions in the open (as opposed to within a plant canopy) is provided by the Campbell–Stokes sunshine recorder. This is a glass sphere which focuses the sun's rays onto a special edaphic factors of the habitat. The role of vegetation in determining soil characteristics should not, however, be underestimated (Chapter 15).

The direct effect of edaphic factors is very great, since the root systems of land plants normally inhabit the soil, and because, with rare exceptions, plants must root where the seed germinates. Both the physical texture and the chemical properties of different soils are important in differentiating vegetation, especially when they are extreme and work in the same general direction as the climatic factors (see p. 198).

The biotic factors of the habitat are due to the organisms which directly affect vegetation. Animals act upon plants in various ways, very largely by eating them or parts of them, but also by carrying pollen and seed, and by manuring and otherwise altering the soil.

Plants themselves, as we have seen in earlier chapters, profoundly affect one another, and the effects which they bring about are sometimes included in the biotic factors of the habitat. This use of the term is correct enough when we are considering the habitats of single species, but not, of course, when we are dealing with a plant community as a whole (see p. 141). Invaders from another community may, however, so change the habitat of the community as eventually to destroy it.

Community habitats change like the vegetation itself. Apart from effects due to the slow secular changes of climate and of physiography, the soil may be constantly changing by erosion, rain-wash, or silting, and in stationary soils by leaching (the washing out of soluble salts). These are all *allogenic factors* (see p. 43). The accumulation of humus, one of the most important of soil changes, is a direct reaction of the plant community itself upon its habitat (*autogenic factor*).

Thus we see that the habitat is very complex, and the result of the interactions of a host of different and varying factors. But we must always remember that their actual effect upon the plants (apart from the grosser direct effects of animals and of physical agents such as wind, snow and rushing water) is essentially resolvable into a relatively few physical and chemical processes of major importance: the effect of light on photosynthesis and growth, the effect of temperature on the chemical reactions in the plant body, the evaporating power of the air on the water in the plant, and the effect of the soil solution and its contained ions on the root hairs and through them on the other tissues of the plant.

CLIMATIC FACTORS

It will be clear from the brief discussion of microclimates that useful and meaningful measurement of climatic factors is not a simple task. Measurement in the open of light intensity, air temperature and humidity, rainfall and wind speed will supply the general features of the local climate, but will not necessarily give reliable information on the climatic conditions

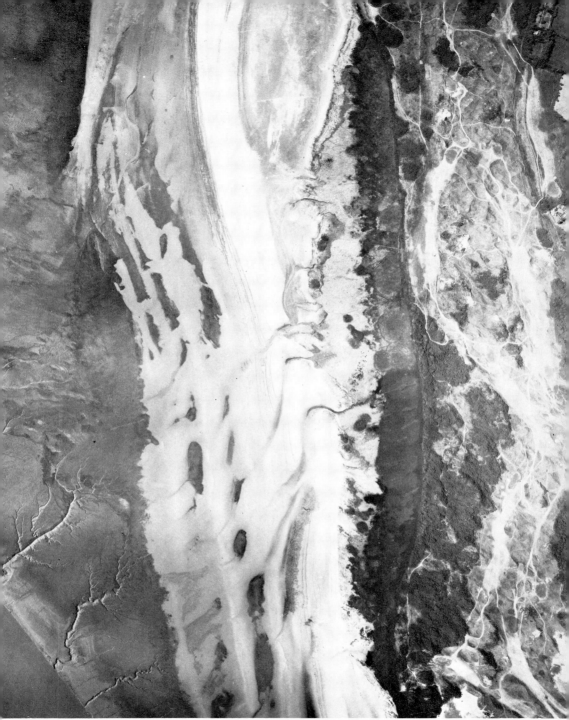

12 Aerial photograph of the coastline at Berrow, Somerset. To the left is part of the estuary of the River Parrett and the 'dendritic' drainage from the mud banks into the main river channel is clear. A vegetated salt marsh flanks an old shore line, to the inland side of which lie sand dunes (with prominent paths and bare areas) which are partly clothed with thickets of sea buckthorn (*Hippophaë rhamnoides*). The darker part of the salt marsh adjoining the dunes supports a mixed vegetation but includes extensive areas of the common reed (*Phragmites communis*) and of sea club-rush (*Scirpus maritimus*); the lighter area, to seaward, is colonized by the cord-grass (*Spartina × townsendii*). A new dune line is building seaward of the salt marsh, inundation of the marsh now (1970) being only at the highest tides (drainage channels, some ending blindly, are visible). Scale approx. 1:7800.

Courtesy of Meridian Airmaps, June 1963

13   Soil profile. Gleyed brown earth, Preston district, Lancashire. The top 25 cm is a dark friable clay loam (A horizon) with some humus at the surface and rusty mottles. The junction with the B horizon is fairly clear, gleyed mottled clay loam extending to the foot of the pit where it merges with the parent material (C horizon) of till. The prismatic to sub-angular blocky structure of the clay is clear.

L. F. Curtis

existing in the vegetation under study. Single observations are rarely of value and comparative measurements should always be made, in space as well as in time. The vertical structure of vegetation is important in determining microclimate and so it is rewarding to make simultaneous measurements of climatic factors at different heights within and above the plant canopy. Diurnal and seasonal variations in such profile measurements supply much information about the climatic conditions actually experienced by plants. The following account of climatic factors outlines important features of their nature and their measurement. Addresses of suppliers of some items of equipment are given in the Appendix.

1. *Light*

Solar radiation, as the ultimate form of energy available for life, is the most fundamental of the climatic factors and its measurement is of considerable importance. Difficulties of precise measurement and valid interpretation are great, but they should not discourage the student from framing and exploring pertinent questions about the light climate of plant communities or the distribution of individual species in relation to light.

Radiation of biological significance can be separated into two classes: solar, or short-wave, radiation with a range of wavelength from about 0.3–3.0 μm (1 μm = 1 micron = 10^{-3} millimetres) and terrestrial, or long-wave, radiation extending from about 3–100 μm. Solar radiation includes those wavelengths which are essential for such plant functions and phenomena as photosynthesis, photoperiodism and phototropism. It is composed of ultra-violet light, the visible spectrum and 'near' infra-red light. An indication of the spectral distribution of solar energy is given in Fig. 14.1a, which shows that the solar energy reaching the earth's surface is greatest in the middle region of the visible spectrum. Terrestrial ('far' infra-red) radiation includes the wavelengths through which bodies lose heat and thus it affects the heat balance of plants. Full discussion of the energy regime of environments and of the interaction of plants and animals with energy exchanges is given by Gates (1962). Problems of measurement and interpretation are considered fully in the British Ecological Society Symposium volume on *Light as an Ecological Factor* (Bainbridge, Evans and Rackham 1966). Further, much information concerning the micro-meteorology of plants and animals is given by Monteith (1973).

Both classes of radiation can be accurately measured only with radiometers, most of which are based on thermopiles (which are a number of thermocouples in series) measuring the heating effect of radiation of all wavelengths. The low output of these instruments must be recorded with sensitive equipment, but simple models are not difficult to construct (see Szeicz 1968). Other, more readily available, devices for measuring light (that is, solar radiation) suffer from the defect of being sensitive only to a

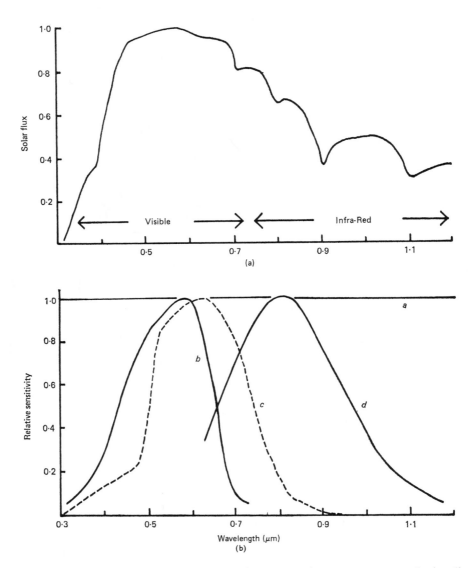

Fig. 14.1. (a) The spectral distribution of solar energy; the curve represents the irradiance of a direct solar beam for a solar elevation of 41° under standard atmospheric conditions normalized to a maximum of 1·0 between 0·56 and 0·60 μm. (Based on data by O. Avaste, H. Moldau and K. S. Schifrin, *Akad. Nauk. Est. SSR Inst. Phys. Astron.* 1962.)

(b) The spectral sensitivity of radiation meters: *a*, thermopile; *b*, selenium (photovoltaic) cell; *c*, cadmium sulphide cell (photoresistor); *d*, silicon solar cell (makers' data).

restricted waveband (Fig. 14.1b), but so long as this drawback is recognized they can be put to good use by the ecologist.

It is important to remember that different organisms, and also different processes within the same organism, are selectively sensitive to the spectral composition of radiation. For example, the human eye is sensitive to the waveband 0·4–0·7 μm with peak sensitivity around 0·55 μm, whereas photosynthesis is 'driven' primarily by radiation in the 0·4–0·5 and 0·6–0·7 μm wavebands and photoperiodism is stimulated by the 0·6–0·7 μm waveband. Thus the ecologist is faced with spectral selectivity both on the part of the organisms which he is studying and the instruments most readily available to him. There should be care in the planning of relevant measurements of radiation and caution in their interpretation.

Two features of the light climate of plant communities have a bearing on its measurement. Firstly the incident light in an open habitat comes from two sources: *direct* sunlight, and *diffuse* skylight which has been reflected or scattered by clouds and dust in the atmosphere and comes from all directions. The relative magnitude of these components varies with the cloudiness, dustiness and humidity of the atmosphere and the angle of the sun. Secondly, the intensity of incident light and its spectral composition vary from minute to minute. This is true in the open with cloud cover, but is especially so within a plant community as the sun moves across the foliage canopy. The most satisfactory conditions for measuring light intensity are cloudless or uniformly overcast skies.

The ability to measure light is not in itself sufficient for biological purposes. It will already be apparent that informative measurements involve comparison of conditions within and outside a plant community, and at different places within the community. They also involve measurement at the most appropriate time of year. For example, as already mentioned (Chapter 3), many woodland herbs produce their leaves and flowers in the 'prevernal aspect', before the foliage of trees and shrubs has developed. Readings should therefore be made during this 'light phase', say in mid-April, as well as in the 'shade phase' at midsummer by which time the leaves of some of the herbs (e.g. *Endymion non-scriptus*) are dying off, although those of others (e.g. *Mercurialis perennis*) persist throughout the summer. Whenever possible, for autecological purposes, measurements should be related to the habitats of particular individuals of the species being studied, and may even take the form of measurements in the plane of particular leaves.

Photoelectric and photochemical methods are available for measuring light intensity. The former are to be preferred since the response to light is usually better defined. Of photochemical methods, the time taken for light-sensitive paper to darken to a standard shade has been used, as has the oxidation of anthracene to di-anthracene. Both methods are unduly sensitive to the ultra-violet (Anderson 1964a). Photoelectric methods include photovoltaic cells (selenium barrier cells) and photoresistors or photoconductive cells (cadmium sulphide cells). An ordinary photographic exposure meter can give a useful indication of the intensity of light reaching or reflected from a habitat, but more precise photovoltaic cells are available and a light meter based on these has the advantage that the recording meter can be remote from the cell, with the possibility of virtually simultaneous readings from several cells. Selenium photovoltaic cells have a response curve fairly similar to that of the human eye. Silicon photovoltaic cells are more sensitive to the near infra-red (Fig. 14.1d). Silicon 'solar cells', developed to convert solar radiation into electrical energy, have the advantage of a higher power output than the conventional design and they are consequently valuable for measuring light at low intensities.

Photovoltaic cells depend on the generation of an e.m.f., which can be measured with a galvanometer, when the receptive surface is illuminated. Useful information concerning the construction of a simple light meter of this type is given by Dowdeswell and Humby (1953). Megatron cells, available as flat plates or discs, are especially convenient for ecological use. 'Potted' cells, sealed in epoxy resin (Araldite), are protected from moisture and may be attached to a handle and employed under water. In open habitats, for the measurement of direct illumination it is desirable to use a diffusing screen (a colour-neutral filter of muslin or nylon gauze) to avoid over-excitation of the receptive surface of the cell, with a consequent loss of sensitivity. The effect of this is to keep the reading on a sensitive part of the range. The extent to which the filter reduces the light intensity can be estimated by comparison with another, calibrated, light meter. If the cell is to be connected to the meter by a long cable for remote readings (as may be desirable for example under thick scrub), the light meter should be calibrated against a standard source with and without the long cable, so that allowance can be made for voltage drop across the cable.

Photoresistors depend for their sensitivity on a change in conductivity on illumination, the resistance falling substantially in the light. They can be used as light meters by measuring the current flowing when a battery is in circuit. If a variable resistance is also included (Fig. 14.2), a wide range of resistance of the cell can be covered and so large differences in light intensity estimated.

The leaves of most plants are orientated such that the blades are at right-angles to the maximum incident light, being held so that they form a closely fitting mosaic. Nevertheless, light from any angle can be used by the plant, but most light meters are relatively insensitive to oblique light. Commercially available photovoltaic cells with 'cosine corrected' diffusing heads are designed to overcome this difficulty. Alternatively just over half of a table-tennis ball cemented over the cell gives some correction (Powell and Heath 1964). A black disc painted onto the top of the ball improves the correction. A different approach is to measure the light reflected from a matt-white surface, screening the light meter in such a way that it can receive light only from the reflecting surface at perpendicular or near-perpendicular incidence (Fig. 14.3). This method was developed by Evans (1956) to study the light climate of woodlands, by integrating the light received over an area of some 1400 cm², rather than merely the light received at the small sensitive area of the meter. Such features as sun flecks can be estimated over the reflecting surface.

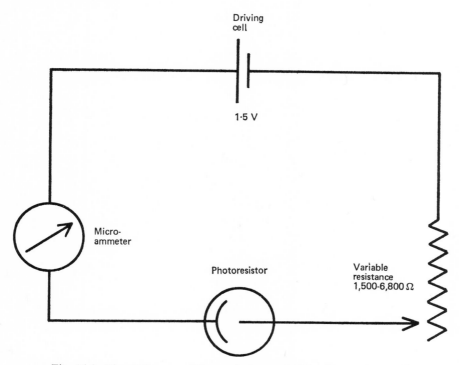

Fig. 14.2. Circuit for use with a cadmium sulphide photoresistor cell.

As the light regime may vary considerably from time to time during the day, and from day to day, being especially dependent on cloud cover, it is usually necessary to take a series of spot readings and obtain an average value. A valuable alternative is to use an integrating photometer which can be left at the site for a substantial period and which gives a measure of the total light received during that period. A simple integrating photometer constructed from a barrier layer photocell can be used, measurements being made of the output of this cell by assessing the transfer of copper between the electrodes, detachable for weighing, of a voltameter (Powell and Heath 1964). The photocells can be made small, light and weather-proof, and separated by a lead from the current integrator if desired, as, paper chart and leaves a burnt trace whenever there is bright sunshine. The chart is changed daily and provides a permanent record of the time and duration of bright sunshine, from which much further information about light conditions can be deduced using, for example, available data on the spectral composition of sunlight. The Campbell–Stokes recorder is the light-measuring instrument most frequently used by local meteorological stations (run by museums, local authorities, schools, etc.) and the daily charts are usually available for consultation.

Hemispherical photography, in which a picture of the full hemisphere of the sky vault is obtained with a 'fish-eye' lens, is a useful, indirect means of assessing the light climate at any place. Astronomical almanacs contain data which enable the apparent course of the sun at any time of the year to be plotted on the photograph and appropriate calculations enable the quantity of incoming radiation to be estimated for all or any part of the sky, taking account of cloud conditions and of such obstructions as buildings or tree canopies (Evans and Coombe 1959; Anderson 1964b). Most fish-eye lenses are extremely expensive, but cheaper models are occasionally available. An advantage of the method is that a single photograph can supply information about light conditions at the site throughout the year.

2. *Temperature of the Air*

A continuous record of air temperature may be obtained with a thermograph, which records temperature on a chart held on a slowly revolving drum. In the absence of a thermograph, maximum and minimum thermometers may be used to give an indication of the temperature extremes of the habitat. In both cases the instruments must be shielded from direct sunlight to avoid errors due to radiative heating of the sensitive parts of the instrument. Shields should be made of materials with poor heat conductive properties and should be covered with a special reflecting white paint or with polished metal foil. It is important to ensure that the shield does not interfere with the free ventilation of the instruments.

Profile measurements of air temperature may be made with suitably shielded bulb thermometers held at the required heights by clamps fixed to a vertical stake or pole. The bulbs of the thermometers should be as far as possible from the disturbing influence of the stake. When cheap thermometers are used, their readings should be calibrated in a water bath over the biologically important range ($0°$–$30°$ C.), as individual thermometers may differ by as much as $1°$ C.

More elaborate, but expensive, methods of temperature measurement include the use of wire resistance thermometers, thermistors and thermocouples, all of which exploit the relationships between temperature and the electrical properties of metals or metal oxides. One advantage of electrical methods is that the temperature can be read or recorded remote from the actual position of the sensitive probe. This can be useful if the temperature of a woodland tree canopy is required or if the presence of an observer would cause undue disturbance, as perhaps in a field of corn. Another advantage is that the probes can be made sufficiently small to record the temperature of leaves, which may under certain conditions differ by several degrees from that of the surrounding air. The chief disadvantage of these methods is that of cost, since they require associated recording instruments, usually of the Wheatstone bridge type. Thermocouples and other very small temperature-sensitive probes are responsive to rapid changes of temperature. This may be an advantage for some studies, but for others it will be found necessary to embed the probe in a body with a larger heat capacity. Conversely thermographs and mercury or alcohol thermometers have an appreciable response lag which makes them unsuitable for measuring temperature in rapidly changing conditions.

3. *Rainfall*
Indirectly this is a factor of prime significance for plants, but it is seldom a factor of *direct* importance. The total rainfall, and especially the distribution of rainfall through the year, is one of the leading features of climate. Sufficiently close figures can usually be obtained from the nearest rainfall station, which can be found on application to the Meteorological Office. Particularly where there is strong land relief, however, notably in mountain regions, the local rainfall may vary very considerably within quite short distances, and additional rainfall records are always valuable and sometimes essential. An automatic rain-gauge is the most useful instrument. This registers the rainfall on a chart and empties itself automatically when the receiver is full. Like the thermograph, it requires a new chart once a week. Non-recording gauges need to be emptied daily for the best results, or precautions must be taken against evaporation of the collected rain. Such gauges may be made simply and effectively out of domestic tins with a polythene funnel and a collecting bottle. The orifice should preferably be 5 in (12·7 cm) in diameter (the British Meteorological Office standard) and the sides of the tin should project above the rim of the funnel for 2 in (5 cm) to minimize splash losses. For meteorological purposes, certain precautions as to location of gauges have to be observed, in order to avoid deviations due to local conditions as, for example, the sheltering effect of a tree; but some of these precautions do not apply to the measurement of rainfall in a given habitat, where local conditions are the important ones which it is desired to record. Nevertheless the greatest source of error in the measurement of rainfall is probably due to air turbulence around the gauge. This can be minimized by surrounding the gauge with a turf wall, or by sinking it so that its orifice is level with the ground. In the latter case precautions must be taken to prevent rain splashing in from the ground.

Rainfall is normally measured in the open, before it reaches vegetation, but for some purposes it may be desirable to follow the fate of rain-water within the plant canopy. A proportion of the rainfall passes straight through the canopy or drips from the canopy to the ground ('throughfall'), some reaches the ground via the plant stems ('stem-flow'), while the remainder evaporates from the aerial parts of the plants, and never reaches the ground ('interception'). In woodland communities it is possible to measure throughfall with several rain-gauges distributed at random over the wood-land floor. These should be placed in new positions at least once a week to avoid over- or under-estimating throughfall because of very local varia-tions. Stem flow may be measured by fixing a spiral gutter around the base of a number of trees, each gutter leading to a collecting bottle at the foot of the tree. Aluminium or zinc sheeting or split polythene tubing makes suitable gutters, and the joint between gutter and stem should be sealed with an appropriate waterproof cement. Interception is estimated by the difference between total rainfall (as measured in the open or above the plant canopy) and the sum of throughfall and stem flow. A similar partition of rainfall in herbaceous communities is less easy to obtain, but with a little ingenuity it is not impossible, at least in communities of tall-growing plants such as bracken (*Pteridium aquilinum*) or rosebay willow-herb (*Epilobium angustifolium*).

4. *Humidity and Evaporating Power of the Air*

This is a very important factor, far more so than rainfall, since it directly affects transpiration (loss of water by evaporation from the aerial parts of the plant), one of the leading processes through which the habitat of a plant is determined.

Atmospheric humidity is commonly measured with a pair of thermo-meters, one with a 'wet', the other with a 'dry' bulb. The 'wet' bulb is surrounded with a muslin sleeve which is either moistened before a measurement is taken or is kept permanently moist by a wick leading from a reservoir of distilled water. Water evaporating from the sleeve lowers the temperature of the wet bulb proportionally to the rate of evaporation, so the difference between the temperatures recorded by the two thermo-meters is a measure of the deficit of water vapour in the air below satura-tion point at the given dry bulb temperature. If the air is saturated with water vapour, the two thermometers record the same temperature. It is important that the wet bulb should be adequately ventilated, for evapora-tion to be unimpeded, and standard instruments, such as the Assmann hygrometer, incorporate a motor-driven fan to draw air over the thermo-meters at a constant rate. It is also important that the wick is kept clean and is thoroughly wetted. Care should be taken to avoid handling it unnecessarily and it should be replaced at frequent intervals by a new, freshly boiled wick (once a week with a continuously operating instru-ment).

The most frequently encountered measure of atmospheric humidity is *relative humidity*, which is defined as the percentage amount of water vapour actually held by the air relative to the amount required for saturation at the same temperature. Since the capacity of air to carry water vapour increases as the temperature rises, a more useful measure of humidity is *saturation deficit*, which is defined as the amount by which the partial pressure of water vapour in the air falls short of the partial pressure at saturation point, whatever the temperature. This measures the actual evaporating power of the air at a given moment, which has more direct relevance to the plant. Both relative humidity and saturation deficit may be calculated from the readings of wet and dry bulb thermometers by reference to hygrometrical tables.

Recording hygrometers (hygrographs) provide a permanent record of humidity over an extended period. Thermohygrographs are available, recording temperature and relative humidity on the same chart. Hygrometers which depend on the property of hair to absorb moisture, with a resultant change in length, require frequent recalibration if reliable measurements are to be obtained. These instruments usually show a lag in response to sudden changes of temperature and humidity, and thus are unreliable guides to short-period changes. If simultaneous readings of temperature and humidity are to be taken from the charts, it is essential to check the alignment of the two pens at the beginning and end of every record.

Atmometers (*evaporimeters*) measure the evaporation of water from an exposed surface and give a useful indication of the total evaporating power of the air at a given spot over any desired period, thus integrating the water vapour content of the air with temperature, wind and the time factor. Unless the air is completely saturated with water vapour, wind raises the evaporating power of the air by constantly removing the saturated air in contact with a wet surface and replacing it by drier air. Two groups of atmometers are available: the open-pan type, in which water is allowed to evaporate from a free surface, and the type in which water evaporates from the surface of a porous material such as paper or porcelain. The second group is in many ways more convenient and adaptable for the ecologist than the first, but with both it must be recognized that the individual properties and the siting of each instrument will have a strong influence on the rate of evaporation and so comparisons between instruments must be made with caution.

The atmometer shown in Fig. 14.4a incorporates a porous pot 'candle' as the evaporating surface. Water lost by evaporation is replaced from the reservoir (a 250 ml flask) via a length of glass tubing which has a side opening at the upper end. The side opening is covered by a length of rubber tubing which acts as a non-return valve, thus preventing water from re-entering the reservoir when the candle is saturated by rain. The candle must

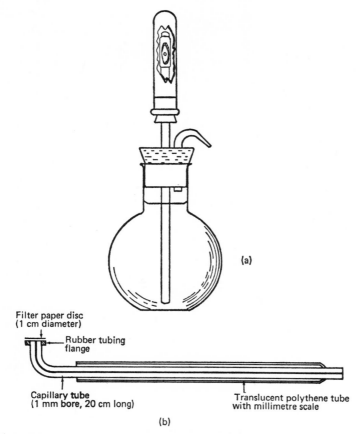

Fig. 14.4. (a) Atmometer for determining the weight of water lost over considerable periods.

(b) A simple atmometer for experiments of short duration. (In part after J. Warren Wilson. *J. Ecol.* 1959, courtesy of British Ecological Society.)

be saturated with water initially and all air bubbles must be excluded from the candle and tubing. Distilled water should be used, and the surface of the candle should be carefully cleaned periodically. The rate of evaporation is determined by weighing the whole instrument at intervals. The size of the reservoir is sufficient to last about a week under field conditions, and so this atmometer is most useful for relatively long-term studies.

A much smaller atmometer, suitable for measuring the rate of evaporation within herbaceous communities, is illustrated in Fig. 14.4b. It consists of a 20 cm length of 1 mm-bore glass capillary tubing which has a right-angle bend near one end. The end of the short arm is ground flat and a sleeve of rubber tubing extends the diameter of the arm to 1 cm. A disc of filter paper is placed on the end of the tube and its upper side acts as the evaporating surface. The water-filled capillary serves as the reservoir and the rate of evaporation is measured by timing the movement of the water–air meniscus in the long arm of the tube. A strip of millimetre graph paper held in a sleeve of transparent polythene tubing provides a moveable graduated scale. The long arm of the tube and the paper disc must be kept in a horizontal position when the instrument is in use. The small capacity of the capillary tube means that the period of observation must be relatively short. A number of these atmometers may be used simultaneously at various heights above ground to obtain profiles of evaporation rates.

5. *Wind*

Violent winds often break off twigs or branches of plants, especially of trees and shrubs, and in very windy situations the tree or shrub may be altogether prevented in this way from growing above a certain height; but the most important general effect of wind is to increase the rate of transpiration by constantly bringing unsaturated air into contact with leaves and young shoots of plants. When a tree such as an oak is exposed to constant and violent winds, mainly from one general direction, not only does the trunk slope away from the wind but new shoots on the windward side are constantly dried and killed off in this way; thus the crown of the tree develops asymmetrically, being directed away from the wind. The same effect is seen in severely wind-cut scrub on exposed coasts and at high altitudes. The drier the air and the higher the velocity of the wind the greater is this effect. In its extreme form it can be seen by the 'blighting' effect of a strong dry wind on the young shoots of all kinds of plants, especially in the spring by an east or north-east wind (the directions from which the driest winds usually come in Britain). Under such conditions the whole of the young leaves and shoots in exposed situations may be killed in a few hours, because the loss of water by transpiration is too rapid to be replaced quickly enough from the roots of the plants.

Even under less extreme conditions, constant exposure to wind, commonly to the prevailing west and south-west winds, though these are more or less laden with moisture, may be of major importance in determining the local distribution of species or communities. Certain trees are well known to the forester to be 'wind-resistant', either because they are better protected against excessive transpiration, or because their twigs and branches are less brittle. For example beech is much more wind-resistant than ash or oak, and oak cannot establish on the exposed south-west sides of woods, while beech can. In this way the succession of establishment of trees in forest development may be greatly modified locally.

Because of decreasing friction with the soil surface the speed of the wind increases very rapidly with increased height above the ground, and the effect of this on vegetation is often exceedingly striking. If a strong wind carries sand particles, these add an erosive effect to the drying effect of the wind itself, the sand being driven against the exposed parts of plants, pitting them, and, if the action is long continued, eventually disintegrating them. Wind-driven frozen snow particles have a similar effect, as may be seen on high mountains near the altitudinal limits of tree vegetation. These last-mentioned effects are greatest at a certain small height from the surface of the ground or snow, for below this height friction with the surface diminishes the speed of the wind so that it cannot carry the particles, and above a certain height only few of the wind-driven particles can rise.

Wind is also important as an agent in the dispersal of seeds or fruits of many plant species and studies of the ecology of such species are likely to profit from an investigation of the relationship between the position of individual plants in the vegetation, wind speed and effective dispersal.

The speed of wind is measured by anemometers, and its direction by wind vanes. Both factors are recorded at first-order meteorological stations. The ecologist who wishes to measure wind for himself can buy or build simple instruments, or resort to even simpler techniques, such as timing the movement or rate of dispersion of small puffs of smoke or soap bubbles. The Beaufort scale, in which the strength of wind is indicated on a 13-point scale, can be estimated subjectively and applied successfully with a little practice. The scale ranges from dead calm (0) to hurricane (12); a gentle breeze (3) just keeps leaves and small twigs in constant motion (10 miles an hour, 16 km/h), and a 'fresh gale' (8) breaks twigs off trees and generally impedes progress (42 miles an hour, 68 km/h). Probably the simplest instrument is the cup anemometer, in which the rotation of three horizontally mounted cups around a vertical spindle either is recorded as the number of revolutions or is converted directly to wind speed. The hot-wire anemometer, in which the rate of cooling of an electrically heated wire is measured, is potentially a more useful instrument for studying the movement of air within plant canopies, but it is more expensive to construct. Because the evaporating effect of wind is integrated with other factors by the atmometer, a detailed study of wind as such is not often required, at least in simple studies, but the importance of wind should not be overlooked.

PHYSIOGRAPHIC FACTORS

It has already been mentioned that strong topographical relief (steep hills and deep valleys) has a profound effect on vegetation, very largely because it produces characteristic 'local climates' (see p. 143). This effect is more marked in less equable regions than Great Britain, whose climate is on the whole rather uniform.

One striking example may be mentioned from southern Europe; the limestone ridge of Sainte Baume, in Provence, which runs east and west, bears on the north and south of its central portion totally different types of vegetation which have not a single species in common. The lower part of the steep northern face, on which fog often forms and which is entirely protected from the midday strength of the Mediterranean sun, bears virgin beech forest, with holly, yew, abundant lichens and a hygrophilous[1] ground vegetation. The south face is occupied by the highly xerophilous[2] Mediterranean 'garigue' vegetation of sparse shrubs and herbs, whose leaves are strongly protected against excessive water loss. This complete contrast is entirely due to the difference of local climates separated only by 300 or 400 feet (90–120 m) of precipitous rock face rising above the beech forest.

Similar differences, though seldom so extreme, are seen everywhere except in the most arid and in the wettest regional climates. In the northern hemisphere, because of differences in insolation, northern slopes usually bear a vegetation adjusted to damper conditions than that on southern slopes, and often bear the same communities as those of southern slopes at a higher altitude. In general the greater the altitude the damper the climate, except where there is a coastal fog belt so that the higher altitudes are drier. In Great Britain such differences are not, on the whole, extreme; but the vegetation, particularly the subordinate vegetation, of a deep sheltered ravine differs considerably from that of a neighbouring exposed slope. On the other hand, when a valley runs in the direction of the prevailing wind it often acts as a funnel and the vegetation of its floor may show the effects of wind more severely than that of the adjoining slopes. The effect of shelter on windy coasts and mountains is very pronounced. It has also been shown, as we have seen (p. 143), that local climate may vary considerably within a short distance (say 50 cm or even 10 cm) on very uneven ground (microclimate); the shelter of a small rock or hillock sometimes makes a difference, for instance to wind effect, that enables a plant or small community to grow which cannot exist outside the sheltered area.

Differences of aspect, and thus of insolation, sometimes affect grassland vegetation markedly, producing on the southern slopes dry, open communities and on the northern slopes tall, moist communities, often with many herbs more characteristic of woodland. In open situations, even small differences in slope and aspect of the ground can affect soil temperatures appreciably, with a concurrent influence on the germination of seeds. A simple home-made clinometer is a useful instrument for measuring the angle of slope of the ground and the angular height of obstructions on the horizon. Such obstructions are generally considered unimportant if they subtend less than 10° with the horizontal. Less than 10 per cent of the diffuse skylight comes from below this angle, and except in winter the sun is higher than 10° for all except short periods after sunrise and before sunset. With compass bearings and angular measurements, the maximum

---

[1] Greek ὑγρός (hugros), moist, and φίλος (philos), loved.
[2] Greek ξερός (xeros), dry.

possible time of direct illumination can be calculated for different times of the year, with the aid of appropriate astronomical data about the apparent movement of the sun. Hemispherical photography can help here (p. 151).

The nature of the underlying rock is often regarded as an edaphic factor, and so it may be; but in so far as it determines topography it is physiographic. Different types of crystalline rocks, hard and soft limestones and sandstones, stratified shales and unstratified clays, and alluvium of different kinds, will produce different topographies, and thus, in conjunction with the climate, affect the physiographic factors. A discussion of such matters belongs to physical (dynamic) geology, some knowledge of which (and the more the better) is usually necessary to the ecologist.

In areas of relatively mature land structure, such as in the Midlands and the south and east of England, the topography is changing very slowly. The land forms are relatively static, and geodynamic changes are not now altering the conditions of vegetation except locally and on a small scale.

It is otherwise in mountain districts and on the sea coast, where geodynamic agents are active. Here the topography may be constantly shifting, destroying the habitats of some communities and creating new habitats. Rock faces, cliffs, steep slopes, and river banks that are being eroded present a special class of habitat in which plants maintain a precarious existence. If the erosion is rapid enough, no species may be able to secure a foothold. Streams cut back into plateaux, increasing the drainage and diminishing the water content of the soil. The eroded material, gravel, sand or clay, or a mixture of these, is brought down by the stream and deposited as silt along its lower course, or creates a delta where the stream flows into a lake, thus providing new habitats for vegetation of different types. Frost, again, breaks up the rocks on flat exposed mountain tops, producing the special type of habitat consisting of more or less loose rocks of various sizes, known by the geologist as 'mountain-top detritus'. Frost and water between them cause the fall of rock fragments from a rock cliff and thus produce screes at its foot—another type of habitat. All these processes change old habitats or create new ones, and present innumerable problems to the ecologist, relating both to succession and to the factors actually at work in determining the vegetation which appears. The geodynamic factors may, as in the case of an evenly eroding rock face, actually maintain a constant type of habitat, and thus a constant type of vegetation, though the individual plants are continually disappearing and being replaced by other individuals of the same species.

A parallel series of problems, though some of them are in detail very different, is provided by the sea coast. Here we have on the one hand sea cliffs composed of different kinds of rock, all of which are being eroded, not only by rain but by the sea, quickly or slowly, and are inhabited by certain species of plants forming various communities. On flat coasts, on the other hand, we have salt marshes, sand dunes and shingle beaches, where new soil is being accumulated, each type of habitat presenting separate successional and strictly ecological problems. Each of the three may show a straightforward development from the first colonization of the new habitat, or this may be modified by continual silting with salt mud, continual covering with blown sand, or accretion of fresh shingle; or the habitat may be destroyed at any stage by tidal or wind erosion and the succession started afresh. Or again, the geodynamic factors of tidal or wind action may balance one another and maintain a condition of equilibrium between accretion and erosion, and thus a constant type of vegetation.

14 Podsol, Exmoor, Somerset. Beneath the acidic grassland vegetation is black amorphous peat (14 cm, A0 horizon), below which is pale grey (eluviated) sandy loam (10 cm) with root mats. Three layers may be seen in the illuvial (B) horizon: a narrow (0·5 cm) wavy iron-pan (arrowed); beneath is silt loam (21 cm) rich in iron sesquioxide; and below this is yellowish silt loam (56 cm) with less iron. The parent material (C horizon) of stony sandy loam is just reached at the bottom of the pit. (The cutting screw of the auger is 23 cm.)

L. F. Curtis

15 The effect of prevention of sheep grazing in Padley Wood, near Sheffield. The low vegetation in the foreground contrasts sharply with that beyond the fence from which sheep have been excluded for about sixteen years. Natural regeneration of trees, notably birch (*Betula* spp.) and oak (*Quercus petraea*) has occurred in the protected area, and there has also been vigorous growth of bilberry (*Vaccinium myrtillus*) and marked changes in the ground flora.

Glyn Woods

Chapter 15

Edaphic Factors: The Soil

Edaphic factors are those due to the soil in which the plant is rooted, and it is usually easy to distinguish between these and climatic factors, though, as we have already seen, the characters of the soil are largely dependent upon climate. For example, the soils of a desert are very different from those of a region of high rainfall well distributed through the year, even if they are derived from rocks lithologically the same.

Relation of Climate and Soil: A Hypothetical Case
Suppose a well-developed bed of limestone of uniform constitution and with a uniform dip is exposed for many miles across country, and that the climate changes steadily as we pass along the strike of the bed. At one end there may be an annual rainfall of 5 inches (125 mm), at the other of 80 inches (2000 mm). Such extreme changes of regional climate within 100 miles (160 km) are well known, though they are not common in western Europe. If our example was in the tropics the low rainfall end would be very arid desert, with an exceedingly sparse vegetation of plants highly tolerant of scarcity of water. The high rainfall end would bear tropical rain forest, the most complex and highly developed vegetation in the world. The soil of the two ends would be totally different, though derived from the same rock. In the desert the rock would be largely bare, and what soil there was would consist of rock particles disintegrated by heat and wind. In the region of high rainfall there would be disintegration of the rock to some depth, and the comparatively deep soil would be covered by and partly mixed with a layer of organic matter, and would constantly retain a considerable amount of moisture.

If the outcrop of limestone was accompanied on each side by parallel uniform beds of very different lithological nature, for instance by sandstone and alluvium respectively, also extending into the two extreme climatic regions, it is probable that at neither end of the climatic scale—neither in the desert nor in the rain forest—would the different rock types show any fundamental difference in the physiognomy (general appearance) of the vegetation. The extreme climatic factors would dominate the situation in each instance in this respect. However, a more detailed study of the vegetation would reveal that the particular assemblage of species present at any one place was determined by a combination of factors including climate, physical nature of the habitat and soil type. In this example the climate can be regarded as the dominating factor in determining the kind of vegetation present but soil type and other factors are important in differentiating the actual communities present within each vegetation formation.

The role of edaphic factors is more clearly seen in temperate latitudes, where, especially if the rainfall is unevenly distributed throughout the year, as it usually is, so that there are wet and dry seasons, the difference of the underlying rocks and of their reactions to the climate and to the vegetation would probably cause a more marked differentiation of the vegetation. With a 20–40 inch (500–1000 mm) rainfall in temperate regions, the limestone soils would tend to be dry and shallow, and the chemical effect of the lime on the soil would also influence the vegetation, while adjoining soils might be almost constantly moist. Here then the edaphic factors would be master factors because they would clearly differentiate the plant communities inhabiting the soils on the different types of rock.

In these examples, however, it would be unwise to over-emphasize the importance of individual environmental factors because they all interact closely with one another (see p. 143).

THE SOIL

The study of the soil as such is now regarded as a separate branch of science known as *pedology* (Greek πέδον, *pĕdon*, the ground, soil). It is only comparatively recently that the immense complexity of the processes that take place in the soil and their profound influence on plants have been at all fully realized.

It is impossible here to do more than give a brief account of some of the features of soil, of different kinds of soil, and of their relations to vegetation, which the field ecologist will encounter. Investigation of the *exact* ways in which different soils affect plants involve some of the more difficult problems of plant physiology, plant biochemistry and physical chemistry.

Definition. The soil may be defined as the superficial covering of the earth's crust, created and continuously affected by surface agencies—heat, frost, wind, and above all water—acting upon the rocks from which the soil is derived and continuously upon the soil itself. Soil is also essentially affected (and sometimes created, as in peat soils) by the vegetation which grows upon it, and it is much influenced by the animals living within it. This continuing action of surface or 'weathering' agencies naturally leads to the development of *stratification*, i.e. the differentiation of distinct layers or *horizons* of soil, from the soil surface down to the unaltered parent material. The formation of clearly distinguishable soil horizons, which together constitute the *soil profile*, is a conspicuous feature of almost all natural undisturbed mature soils, which have virtually reached a state of equilibrium with soil-forming agencies.

To the plant ecologist the soil is the surface material of the earth in which plants grow, containing the underground parts of higher plants as well as algae, fungi and bacteria—often collectively known as the *soil flora*. There is also a *soil fauna*, which may be very rich and varied, ranging from Protozoa to different kinds of worms, insects and other invertebrate animals (Jackson and Raw 1966) and including small mammals such as moles, mice and voles, which spend a great deal of their lives below ground. An indication of the abundance of various micro-organisms which may be found in fertile agricultural soil is shown below (after Burges 1958):

	Number per gram of soil
Bacteria	250,000,000
Actinomycetes	700,000
Fungi	400,000
Algae	50,000
Protozoa	30,000

These soil organisms, surrounded by the mineral and the dead organic constituents of the soil and by the air and water it contains, form a distinct microcosm at any spot with its own atmosphere and water supply in which chemical, physical and biological changes are constantly taking place. The different parts of this microcosm, especially in a mature and stable soil which has reached a state of approximate equilibrium, show a very complex system of mutual relationships which contribute to and accentuate the very heterogeneous nature of soil at both the micro- and macro-scale.

THE FRAMEWORK

1. *Chemical Characteristics*

The major part of most soils is the inorganic fraction produced by the disintegration (weathering), by both physical and chemical agencies, of the parent material which may be hard rocks—igneous or metamorphic, or sedimentary grits, sandstones or limestones—or softer shales, clays and alluvia. Besides this inorganic fraction, organic material is almost always present. Dependent on the particular type of rock from which the soil is derived, the nature of the soil differs both chemically and physically. There are, however, three main inorganic constituents which form the basis of most soils; these are complex alumino-silicates, silica and calcium carbonate. The alumino-silicates are derived from such rock-forming minerals as the felspars, hornblende, augite and mica of igneous and metamorphic rocks. The more resistant of these minerals pass into sedimentary rocks by erosion, transport and redeposition, usually through the agency of water, while the less resistant minerals are dissolved away. The silica comes largely from the quartz of acidic igneous rocks and also is the major constituent of sedimentary rocks such as gritstones and sandstones. Calcium carbonate, on the other hand, comes mainly from limestones originally formed in seas of previous geological eras by such organisms as corals, foraminifera and calcareous algae.

With the alumino-silicates are associated various bases such as calcium, magnesium, potassium and sodium which, except for sodium, are important in plant nutrition. Besides these elements, phosphorus and sulphur, usually in the form of phosphates and sulphates or the corresponding acids, are always present and are essential to plant life. Because these nutrient elements are found in many of the alumino-silicate minerals that are relatively quickly weathered and hence only sparsely represented in most sedimentary rocks, soils which are derived from sedimentary rocks—such as sandstones—are often infertile (poor in plant nutrients) in comparison with soils derived directly from igneous rocks. Iron salts are nearly always present—iron is an essential element in the formation of chlorophyll—and the oxidation from ferrous to ferric salts during weathering gives the brown or red colour of many soils. Nitrogen, which is an essential constituent of plant proteins, is mainly derived from decaying organic matter, but a major ultimate source of nitrogen in the soil is the fixation of the free nitrogen of the air by micro-organisms. In most soils there are small amounts of elements which are needed in minute quantities for the healthy growth of plants; these *trace* elements are boron, manganese, copper, zinc and molybdenum. Other elements (e.g. aluminium, chromium, nickel, lead), not required by plants, also occur and are frequently absorbed. At very low concentrations these have little effect on plant growth, but even at moderate concentrations some are extremely toxic, as also are certain of the trace elements such as copper and zinc.

The process of chemical weathering in the soil breaks down the complex alumino-silicates, largely by hydrolysis, and the bases are removed in solution, while a part of the silica is separated from the complex silicates. From the soluble and colloidal products of weathering the clay mineral particles are formed by crystallization. These secondary clay minerals are important to soil fertility because they are able to adsorb the positively charged metallic ions (cations) of bases such as calcium, magnesium and potassium from solution and exchange the adsorbed ions for other ions, i.e. they show properties of *ion exchange*. The extent to which cations can be exchanged in the soil may be measured, and is referred to as the *cation exchange* or *base exchange capacity*. The clay minerals thus act as a reservoir of mineral ions which are readily available to plant roots. The extent to which the exchange sites are occupied by cations of bases rather than hydrogen ions (the *base saturation*) provides a measure of how far the weathering of parent material and the release of cations of bases keep pace with loss of such ions in drainage water (see Fig. 18.1, p. 204).

The clay-mineral particles are very small—they are not visible with the light microscope—and are often present as colloidal material. This colloidal clay, together with the colloidal humus derived from the decomposition of organic matter, forms a colloidal complex called the *weathering complex* or *clay–humus complex*, the components of which are so intimately associated that it is difficult to separate them without materially affecting their physical and chemical properties.

Calcium is the dominant exchangeable cation in most soils and when present in quantity it imparts physical and chemical stability to the clay–humus complex, aiding the aggregation of the fine colloidal particles into compound particles and thus giving a granular or 'crumb' structure to the very fine-grained clays which, without calcium, are unfavourable to many forms of life.

While the calcium ions are thus the great stabilizing agent in soil, the free hydrogen ions, derived from the ionization of acids, promote chemical change. These are the active agents in the chemical action of acids, and in soil are derived mainly from the carbonic acid dissolved in soil water, from the organic acids of humus, and from other acids produced as a result of chemical changes. The concentration of hydrogen ions in a solution is measured by what is called its pH value. pH is the electrical potential (p) of the solution and is expressed as the negative of the logarithm (to the base 10) of the hydrogen ion concentration, i.e. the number of free hydrogen ions in gram-equivalents per litre (pH $= -\log [H^+]$). Thus pH 7 means 10^{-7} (i.e. 0·0000001) gram-equivalents of hydrogen ions per litre and this expresses the number in pure water, derived from the ionization of water molecules. When an acid is added, some of its molecules are ionized and the number of free hydrogen ions is increased. A pH value of 6, for example, means a concentration of 10^{-6} (i.e. 0·000001) hydrogen ions, ten times that in pure water. A pH value of 3 means 10^{-3} or 0·001, ten thousand times that of pure water. Thus the lower the pH value (negative index) the higher the acidity, and a tenfold change in hydrogen ion concentration is represented by a pH difference of one unit. pH values above 7 indicate alkalinity of the solution.

There are various ways of measuring the soil reaction or pH of a soil. By far the best is direct determination of the electrical potential, using glass electrodes and a potentiometer or pH meter. Another method is the use of colorimetric pH indicators which give useful results when employed with the proper precautions. These depend on the colour changes of certain organic dyes whose colours are known to change in definite ways according to the pH of the solution to which they are added. A selection of these dyes (colour indicators), each of which assumes a certain range of colours through a corresponding range of known acidities and alkalinities, is used for this purpose. A full description of the various techniques for determining soil reaction will be found in M. L. Jackson's *Soil Chemical Analysis* (1962), Chapter 3 (see also Hesse 1971).

Most British soils, except limestone soils saturated with calcium carbonate, those derived from certain basic igneous rocks and saline soils, are more or less acidic in reaction, even if the rock from which they were derived was alkaline. Percolating rain-water, containing excess of hydrogen ions, is active in 'leaching' (washing out) bases from the upper layers and the results of this process are particularly marked in the west and north-west of Britain where the rainfall is heavy and the rocks are mostly poor in bases initially. It is said that the *average* pH value of northern English soils under natural vegetation is about 5, which means that the average soil is distinctly, though not excessively, acidic.

It must not, however, be supposed that because a soil is distinctly acidic in reaction that it is necessarily poor in the cations of bases. Such a soil may be quite lacking in calcium carbonate, to which an alkaline soil reaction is usually due, and yet quite rich in calcium ions adsorbed by the colloids of the clay–humus complex; that is, such soils have a relatively high percentage base saturation (see p. 164). For example, it has been shown that many of the loam soils of the Chiltern plateau, which are decidedly acidic with a pH of between 4 and 5, have a fairly high 'base status', and they support finely grown beechwood and other luxuriant vegetation. On the other hand extremely acidic soils, with a pH value of between 3 and 4, are in fact very poor in bases (low base saturation) and support only a specialized (calcifuge) natural vegetation of plants such as ling (*Calluna vulgaris*), mat grass (*Nardus stricta*) and wavy hair grass (*Deschampsia flexuosa*). These soils require the addition of lime as the first step in rendering them agriculturally fertile. The infertility of these acidic soils is, however, related to the organic fraction of the soils (see p. 170).

2. *Physical Characteristics of Soils: Soil Texture*

The nature and size of the mineral particles constituting the inorganic framework of soils not only influence the chemical processes, but directly determine the physical nature of a soil and its effect upon plants. The *texture* of a soil depends primarily on the sizes of its mineral particles, and this feature is of great importance to plants because it controls aeration, water-holding capacity and the ease with which water can drain through the soil. The proportion of particles of different sizes present in a soil is estimated by the procedure known as *mechanical analysis*, in which the *fractions* of the soil whose particles lie between different limits of size are determined. To these fractions the common names of well-known types of soil are applied, e.g. gravel, sand, silt and clay. The U.S. Department of Agriculture classification of particle sizes is shown in the table (but other standards are sometimes used in which the size ranges of the sand and silt categories are slightly different).

	Size of particle (mm diameter)
Gravel (and stones)	>2
Coarse sand	2–0·5
Medium sand	0·5–0·25
Fine sand	0·25–0·05
Silt	0·05–0·002
Clay	<0·002

All soils in fact contain particles belonging to more than one of these categories, the soil itself being named, by textural class, after the fraction which is preponderant (see Fig. 15.1). When soil consists of a balanced mixture of particles of widely different sizes with an adequate humus content it is called a *loam*.

In mechanical analysis, weighed samples of soil, oven-dried at 100° C., are treated to remove all calcium carbonate and organic matter. The mineral residue is then normally crushed and subjected to a series of sievings with sieves of differing mesh sizes and the proportion of the soil retained

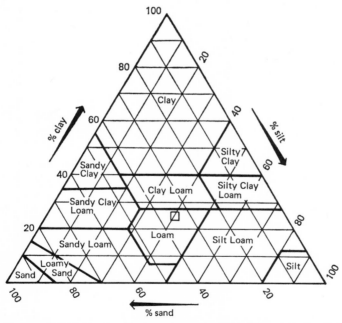

Fig. 15.1. Textural classes of soils based on the U.S. Department of Agriculture classification. The textural classes are shown within the triangle with respect to their percentage composition of the three components: sand (2·0–0·05 mm), silt (0·05–0·002 mm), and clay (below 0·002 mm). The soil shown by □ contains 35 per cent sand, 40 per cent silt and 25 per cent clay and is by texture a loam.

by each sieve size is determined. In some procedures the finer fractions are washed through the sieves. The exact method used may vary in detail, for example in the way in which the organic matter is removed, but during mechanical analysis all compound particles or 'soil crumbs' are broken up into their constituent particles.

A rough qualitative analysis of a sample of soil can be made by shaking up about 10 grams of soil in a large jar of distilled water. The sand fractions sink very rapidly, the silt in the course of a few hours, while the fine clay remains suspended in the water indefinitely. After the sand has sunk the turbid liquid may be carefully decanted into other jars which are then filled with distilled water and left standing. The rate of settling and the final turbidity of the water will then give a fairly good indication of the proportions of the silt and clay fractions. A pinch of washing soda added to the water aids dispersion and thus separation of the particles. This method, involving the rate of settling, is the basis for the separation of the finer fractions in the quantitative procedure of mechanical analysis. An experienced pedologist can also estimate the textural class of a soil by its feel by means of such criteria as are given on p. 184.

Characteristics of Soils of Different Textures. Gravels are soils with a large proportion of particles more than 2 mm in diameter, usually mixed with sand and also some finer particles. Aeration and percolation of water are extremely free, but often gravels are unfavourable for plants because of the likelihood of drought and the frequent deficiency of mineral nutrients.

Sandy soils have a preponderance of particles between 2 mm and 0·02 mm. The particles are typically, but not exclusively, of silica (SiO_2). Percolation of water and aeration are free, but the water-holding capacity (in the absence of abundant humus) and the capillary rise of water are slight, because the soil pores (the spaces between the particles) are large. Hence sands are dry soils unless the ground water is high, and are warm, 'early' soils because, owing to their dryness, they warm up quickly in spring. They are light and easy to work but typically poor in nutrients because of the deficiency in the finer particles with which bases are associated, and because the free percolation of rain-water leads to very thorough leaching. For this reason they are easily 'podsolized' (see p. 178) and quickly become acidic.

In oceanic and suboceanic temperate climates, sandy soils often bear heath vegetation or heathy woodland, characteristically of birch and pine, with a tendency to the formation of raw humus (*mor*, see p. 172). Sandstones with sufficient of the finer fractions to maintain their base status develop the 'Brown Earth profile' with mild humus (*mull*, see p. 172) and support broad-leaved deciduous forest, though coarse sand grains may be abundant in the soil and affect the nature of the subordinate vegetation. Deforested soils of this kind tend to podsolize and develop heath, though they make good agricultural soils for certain crops. Coastal blown sands support a specialized vegetation able to withstand their mobility (p. 68). On stabilization they resemble sandy soils derived from coarse sandstones and tend, with long-continued leaching, to develop heath on their inland margins.

Silt soils are intermediate between sands and clays in the size of their particles. The name is derived from the prevalent texture of alluvial soils (silts) laid down on the flood plains of rivers. Where such soils contain a moderately high admixture of clay they are usually fertile and productive: silt soils often have considerable water-holding capacity, while percolation, aeration and capillary rise of water are fairly free. However, the management of silt soils for agriculture is difficult and conversion to arable cultivation often inadvisable. In Britain these soils probably originally bore alder and oak woodland, but have mostly been converted to meadowland.

Clays are soils in which the particles below 0·002 mm in diameter, typically of hydrated alumino-silicates, are numerous enough to give character to the soil. Any soil with a proportion of from 30 to 40 per cent or more of these fine particles would be called a clay soil. Clay soils have all the qualities contrasting with those of sand: percolation of water is very slow or almost nil, aeration defective, and water-holding capacity very high. Clay soils are wet (often waterlogged for long periods in winter and early spring), heavy, difficult to work, cold and 'late' because they warm up slowly owing to the high water content. Under continued drought the clay colloid shrinks, cracks, and eventually 'bakes' hard. These characters make clay soils physically unfavourable to many plants, and root systems tend to be shallow because of poor aeration at greater depths. This effect can be well seen on grassland on clay, where the grasses are all shallow-rooting species, and in woods on clay, where the roots of shrubs and trees tend to be concentrated in the 'improved' surface soil, failing to penetrate the unaltered clay below, while in a wood on sand the roots of the woody plants are not so restricted to the surface soil and reach a greater depth. On the other hand clay soils are often chemically favourable to plants because they may be rich in bases associated with the clay-mineral particles. Clays are, however, sometimes deficient in essential nutrients, e.g. particularly phosphates, and occasionally in bases also. Clays are much improved by an abundance of mild humus (mull) which 'opens' and lightens the soil, and by calcium carbonate which flocculates the clay colloids so as to form 'crumb structures', in this way leading to better aeration and freer movement of water.

Clay soils in the Midlands, south and east of England support typical damp oakwood. After deforestation, they have been largely converted to 'permanent grass', which is, however, unprofitable when not carefully tended, as this soon reverts to coarse grasses, rushes and scrub vegetation.

Loams, consisting of a mixture of particles of different sizes in which no one fraction predominates, are the most favourable soils for the great majority of plants because they tend to combine the good qualities of the extreme types. Thus the clay and humus fractions give consistency and water-holding power and supply plant nutrients, the particles of medium size permit the capillary rise of water, while the sand particles facilitate aeration. The constitution of loams has a wide range according to the relative preponderance of one fraction or another: thus we have 'heavy (clay) loams', 'medium loams', and 'light (sandy) loams'. With an adequate supply of bases, especially calcium, plenty of mild humus and good water supply and drainage, medium loams are the ideal soils for all plants except highly specialized types suited to extreme edaphic conditions.

Limestone Soils. Soils derived directly from limestones form a class apart. A relatively pure limestone, such as the chalk (often 90 per cent, and in some cases nearly 100 per cent calcium carbonate), weathers by chemical solution of the carbonate, not primarily by mechanical erosion, though this may contribute where the rock is exposed in cliffs and scars. For the most part, gradual solution of the rock takes place below the carpet of vegetation and surface organic matter through the action of percolating water containing carbonic acid (largely derived from carbon dioxide produced by the respiration of plant roots and soil organisms). The result is the formation of a shallow soil consisting of the scanty residue of insoluble mineral particles and of surface humus, the whole saturated with calcium. This is the type of weathering which occurs on the grass-covered slopes of the chalk downs and of the older limestone hills of the north and west, and also below the leaf litter and abundant humus of the beech 'hangers' on chalk escarpments and valley sides. Such shallow limestone soils are dry ones, because the percolating rain-water quickly escapes through the fissures of the rock below. They support a herbaceous vegetation of plants which can tolerate drought, including species ('calcicoles', see p. 190) which flourish on alkaline soils and cannot tolerate acidic conditions (see also p. 181 under *rendzinas*).

Deeper soils are also found overlying limestone but often in such cases the soil material is derived not from the underlying rock type, but from superficial deposits. Examples are the 'clay-with-flints' soils of the plateau areas of the North Downs (see p. 183) and the soils of the Derbyshire Carboniferous limestone which contain loessic (i.e. wind blown) material.

3. *The Organic Fraction of the Soil*
Almost all soils contain organic material derived from the disintegration of plants or parts of plants such as dead leaves, roots and rhizomes, with a small addition from animal excreta or dead bodies. The animal contribution is generally insignificant compared with the great bulk of plant material, apart from highly exceptional soils such as those, largely formed

of birds' droppings (guano), which accumulate below maritime 'bird cliffs' or 'bird rocks' inhabited by thousands of sea birds. A distinction is often made between the organic matter of soil which still has some obvious structure and the amorphous (structureless) organic matter in soil. The term *humus* has been variously used, sometimes for the whole complex of disintegrating and decaying organic material in or on the soil and sometimes only for the brown amorphous substance which is ultimately formed as a result of the processes of disintegration and chemical change of the organic debris. Here it will be used in the latter sense. The soils of habitats such as deserts, in which the vegetation is always sparse, and new soils, freshly formed from inorganic material, contain the least organic matter, whereas mature, well-vegetated soils contain the most. Dead leaves and stems lying on the surface of the soil are distinguished as *litter* before they disintegrate. The amount of litter arising from leaf fall in autumn in woodland, for example, is considerable; in a woodland community the dry weight of the litter from the autumn leaf fall may be about 1 ton per acre (3000 kg/ha).

In soils newly formed from rock or alluvium, organic matter begins to accumulate as soon as plants of any kind colonize it and a certain amount may also be added by wind-borne particles of organic material. In many soils, especially those which are calcareous, earthworms are important agents in incorporating dead leaves and stems into the soil. They drag material down into their burrows and constantly pass large quantities of organic matter through their bodies, disintegrating and partially digesting it in the process. Disintegration is aided by many other small soil animals, and also by soil fungi and bacteria, the proportion of animal faeces in the organic matter increasing as the material passes through successive animals. Eventually the disintegrated and decomposed organic substance is totally broken down by bacteria of different kinds into carbon dioxide, water and mineral salts. Cations yielded by these inorganic salts, mainly calcium, magnesium and potassium, and anions containing nitrogen, phosphorus and sulphur, are essential plant nutrients and their release from complex organic residues by soil micro-organisms is an important feature in maintaining soil fertility (see Fig. 18.1, p. 204). A most important chemical soil process as regards the nutrition of higher plants is *nitrification*, the conversion of ammonium salts into nitrites and then into nitrates, the last process being carried out by the *nitrifying bacteria*. As the great majority of green plants absorb their nitrogen chiefly in the form of nitrates the activity of nitrifying bacteria is of particular ecological importance. The various processes noted above take place most freely and rapidly in soils with a fairly high 'base status' (i.e. fairly high level of cations such as calcium and magnesium) and a moderate water content, in the presence of plenty of free oxygen and at a moderately high temperature, i.e. in rich, moist, warm, well-aerated soils. This is because these conditions are most favourable to the life of the organisms which carry out disintegration and decomposition, from earthworms down to the aerobic nitrifying bacteria.

The kind of humus formed where the processes just described occur freely is known as *mull* (mild humus). Under these conditions the organic turn-over is quick when the temperature is sufficiently high. Humus is formed in quantity, and rapidly, because the favourable conditions for the growth of plants give an abundant supply of plant material, and this is rapidly de-composed because of the favourable conditions for the activity of the soil organisms. In this way an ample supply of the ions of the mineral salts which had been locked up in the plant tissues is set free and made available to plants again. Mull is therefore the characteristic humus type of fertile soils with high base status. The humus is well incorporated in the soil and becomes combined with the compound particles of colloid clay to form the clay–humus complex.

At the other extreme is the type of humus known as *mor*, formed in less favourable conditions under which the organic matter tends to accumulate and becomes very acidic in reaction. There are several conditions leading to the production of mor. Firstly it tends to be formed in soils derived from rocks which are very poor in bases, such as many of the siliceous rocks of the north and west of Britain and some of the sandy soils of the south and east. Secondly it is formed especially in a cold damp climate where the conditions are unfavourable for the active life of many of the mull-produc-ing soil organisms. Thirdly, high rainfall leads to thorough leaching of the surface layers of soil, carrying down the soluble mineral salts to lower layers, especially quickly on highly permeable soils, and thus increasing the poverty of the upper layers and leading to the formation of mor. Where one or more of these conditions prevails the formation of humus from litter is slowed down and partly decomposed litter tends to accumulate, becom-ing highly acidic because the bases contained in the decomposing plant substances are leached out and the organic acids are not neutralized. The excess of hydrogen ions renders the organic substances extremely mobile, and heavy rain carries them down to lower levels, especially in a permeable soil. The solubility and consequent mobility of mor humus are conspicu-ously shown by the brown colour of moorland streams draining from acidic peaty soils, especially when the streams are in spate. In the hilly and mountainous regions of the west and north of Britain all the conditions leading to the production of mor—siliceous rocks poor in bases, a cool damp climate, and heavy rainfall—are frequently combined, so that mor soils are the prevailing type over wide areas.

The soil fauna and flora of mor are very different from those of mull. Earthworms and other invertebrates active in the formation of mull are fewer or absent from acidic mor soils and litter fungi predominate instead of the wide range of bacteria found in mull. The absence of nitrifying bacteria in mor soils means that no nitrates are produced and ammonium compounds are the main form of combined nitrogen available for the nutrition of higher plants. Mor soils with a complete absence of nitrates were found by Pearsall (1938, 1971) to have a pH value of below 3·8, which appears to be a critical limit separating mor from mull in several soil series; however, studies have shown that all transitional types exist between the extremes of mull and mor and that the factors determining the formation of one or the other are not clear cut. The deficiency of bases characteristic of mor soils can be easily tested in the field by shaking up a sample of the soil with Comber's reagent (see p. 186).

Soil Structure. In a natural soil the individual particles tend to become aggregated and to some extent cemented together to form compound aggregates. Humus, inorganic salts and mucilages produced by the soil fauna all contribute in the formation of soil aggregates. These compound particles are known variously as aggregates, crumbs and peds. The structure of a soil refers to the extent to which aggregates are developed and to the size and shape of the aggregates present. The description of soil structure is usually qualitative by use of terms such as weak, moderate and strong to describe the degree of aggregate development and the following terms to describe the type of aggregates present:

Platy, where the vertical axis is shorter than the horizontal.
Prismatic, where the vertical axis is longer than the horizontal.
Blocky, where the vertical and horizontal axes are approximately of the same size and rather angular (see Plate 13).
Granular, when the aggregates are rounded and rather small (up to 1 cm).

The stability of the soil aggregates is also of great importance, particularly in relation to treading by animals and exposure to impact by heavy rain; should the aggregate structure break down, the soil is liable to become 'puddled' (as around farm gateways) and consequently, owing to compaction and poor aeration, unsuitable for plant growth. Plants, especially those with fibrous root systems such as grasses, are important in making, improving and maintaining soil structure. In general, the longer an area is left as grassland the greater will be the proportion of the topsoil in larger and more stable soil aggregates. A soil with a good crumb or aggregate structure is well drained, well aerated and provides an excellent 'tilth' essential for good seed germination and the early growth of seedlings.

Water Content of Soils. The amount of water contained in a given soil at any moment depends upon several factors. We have seen that the water relations of soils of different textures are very different, the amount of water retained by the soil depending on the size of the soil particles and the amount of organic matter present. Water percolates readily through gravels and sands, which are mainly composed of coarse particles, because of the large air spaces between the particles. The smaller the particles, and consequently the air spaces, the slower the rate of percolation and also the longer the soil takes to dry. Very small particles such as those of clay also hold water very strongly by surface forces so that, while a clay soil may contain a great deal of water, only some of that water can be utilized by higher plants. Humus is also a very important water-holding constituent and soils rich in humus retain a considerable amount of water, which they absorb and hold like a sponge for a long time, even in relatively dry air. A soil which has drained freely after rain is said, in terms of moisture content, to be at *field capacity*. At field capacity sandy soils contain much less water (total water content) than clay soils but a greater proportion of that water is held in a form which is available to plants.

In considering the supply of water to the soil we must distinguish between ground water and water coming directly from rainfall or held in the surface layers. *Ground water* is water derived from a more or less permanent 'water table'. In lakes and rivers this is above the soil surface, in a marsh at or very close to the surface, and in many alluvial soils only a short distance below the surface. Underground water may be close to the soil surface when there are underground streams or bodies of water held up by an impermeable stratum below. For example, a thin surface stratum of sand or gravel may be kept constantly moist because it lies over an impervious clay stratum just beneath it. Very often, however, the water table lies so far below the surface that it has no effect on the surface soil, which then depends for its water content entirely on direct precipitation. After heavy rain this increases greatly, and the surface layers become saturated. The excess above field capacity runs off or percolates through to lower levels. Evaporation from the surface of the soil is not rapid in a damp atmosphere and it ceases in a saturated atmosphere. During a long spell of hot weather or of drying winds, however, when the saturation deficit of the air is high, evaporation decreases the water content of the surface layers of the soil to (and beyond) a point at which plants wilt, the soil ultimately becoming 'air-dry'. Wilting occurs because the plants are no longer able to extract sufficient water from the soil to make good the transpiration losses from the leaves. A close cover of vegetation may greatly decrease direct evaporation from the soil surface, but the soil may lose water more rapidly and from a greater depth by absorption through the roots of the plants and transpiration from the more extensive aggregate surface of the leaves than directly by evaporation from the soil surface. Consequently a dense carpet of vegetation tends to dry out the soil.

ZONAL (CLIMATIC) SOILS IN BRITAIN

Brown Earths and Podsols
We have seen that the cool damp climate of the north and west of Britain

is one important factor leading to the formation of mor soils, while the warmer and drier climate of the south and east is more favourable to the production of mull. This contrast is an example of the differentiation of soils by climate which has led to the recognition of what pedologists call *world groups* of soils, dependent primarily on climate. Such soils are also referred to as *zonal soils* so named because they occur over large areas or zones of the world. The two world groups represented in Britain are called respectively *Brown Earths* (or Brown Forest soils), characterized by relatively high base status, the presence of nitrates and the formation of mull, and *Podsols*, characterized by low base status, the absence of nitrates and the formation of mor. The brown colour of Brown Earth is due to hydrated ferric oxide derived from iron-containing alumino-silicates in the parent rock and liberated in the weathering complex. 'Podsol' is a Russian word meaning 'ash', from the pale ash-colour which often marks the layer immediately below the dark surface mor of the typical podsol (see Fig. 15.4).

We saw at the outset that stratification, due to the continuous action of surface agencies, is a feature of nearly all mature natural soils. This stratification gradually develops in a newly formed soil until, in a mature soil, the whole soil profile, down to the subsoil or parent rock, has attained a condition of equilibrium. In regions of moderate and high rainfall the primary agent of stratification is the percolating rain-water and its interaction with soil constituents, both mineral and organic. The descending water has two main effects. Firstly it washes down fine particles from the surface to lower levels, and secondly it dissolves soluble salts and carries their ions downward frequently redepositing salts at a lower level. The result of these actions of percolating water is the impoverishment of the surface layer of the soil in fine particles and in bases, and the accumulation of these in lower layers (Fig. 15.2). Thus we can distinguish in mature soils an impoverished leached *eluvial layer* above from a layer of deposition and accumulation below, the *illuvial layer*; these are generally known by pedologists as the A and B *horizons*. The upper regions are constantly being leached of their soluble components, and fine soil particles, such as clay, are moved downwards, suspended in soil water, by mechanical eluviation. The A and B 'horizons' are well developed in Brown Earths and are even more sharply defined and conspicuous in Podsols (see Plates 13, 14 and Figs. 15.3, 15.4).

Brown Earths are best studied in natural deciduous woodland which has been undisturbed for a long time. They are most typically developed on a clay or loam subsoil (Plate 13, Fig. 15.3), but are also formed on sandstones and limestones containing a large clay or silt fraction. The profile of a typical Brown Earth is free from calcium carbonate, any that was derived from the parent rock being leached out by percolating rain-water and either ionized in the soil water, the calcium ions entering the clay–humus

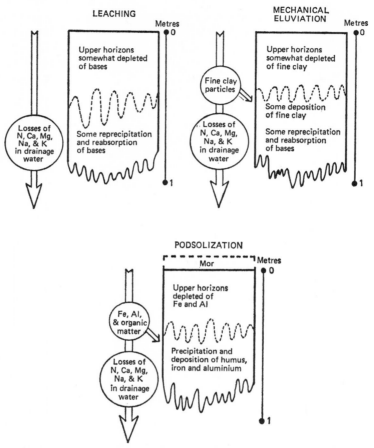

Fig. 15.2. Summary of the processes connected with the downward movement of water in the soil. Such processes are most intense in areas of high rainfall and free drainage. (After C. P. Burnham and D. Mackney, 1964, *Field Studies*, courtesy of Field Studies Council.)

complex, or carried down to lower levels together with the finer clay particles. The whole profile is usually somewhat acidic with a gradient of decreasing acidity and increasing base content downwards. The A horizon (eluvial) has a substantial proportion of finely divided mull well-incorporated with the mineral constituents, and a relatively high base status, the cations of bases, including calcium, being held in the crumb structure of the weathering complex. The A horizon is relatively well aerated and has a rich microflora and fauna, with the presence of earthworms. The B horizon is often not sharply delimited and gradually merges into the unweathered parent material. Sometimes the surface layers have a higher pH and a higher base content than the lower, owing apparently to the upward movement of mineral nutrients in the plants themselves, whence they pass to the humus through leaf fall, or by the capillary rise of soil water in dry weather. There is evidence of an alternating movement of water in many of these soils, downwards in wet weather and upwards in dry weather, when evaporation exceeds precipitation. This tends to maintain the stability of the Brown Earth profile and to prevent podsolization, which is associated with a 'one way' (downward) movement of water.

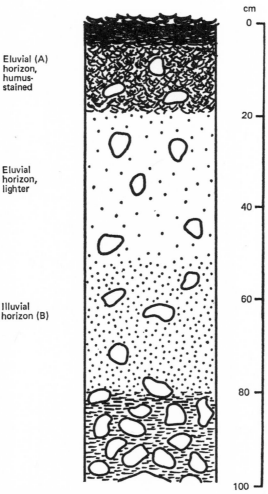

Fig. 15.3. Brown Earth profile. The section is from beechwood on the Chiltern plateau; the parent material is clay-with-flints. The eluvial (A) horizon is humus-stained above; the illuvial (B) horizon is not sharply marked but shows increasing fineness of texture downwards. (Adapted from A. S. Watt, *J. Ecol.* 1934*b*, courtesy of British Ecological Society.)

Most of the good agricultural land of Britain is of the Brown Earth type, but ploughing of course destroys the stratification and tends to make the soil homogeneous on the macro-scale to the depth to which the plough reaches. Manuring enhances the humus content and helps to maintain the base status, since the plant material, which in a woodland is constantly supplied to the soil through leaf fall, is removed with the crop, together with its contained mineral nutrients. Fertilizers maintain base status without replenishing organic matter.

Podsols, with their mor humus, are characteristic, as we have seen, of the cool wet climates of the west and north, where precipitation exceeds evaporation for most of the year, and consequently the downward movement of water is on the average much stronger than the upward. Under such conditions leaching and eluviation are extreme and podsolization occurs (see Fig. 15.2). The prevailing lower temperatures check the disintegration of plant debris and a layer of highly acidic raw humus (mor) tends to form on the surface of the soil. This is called the A0 horizon, since it is an added stratum above the mineral soil (Plate 14, Fig. 15.4). Below the surface mor is the A horizon proper, very poor in bases, often grey or even white in colour, but sometimes stained chocolate brown by the humus substances carried down from A0 by the acidified percolating water. Then we come to the illuvial (B) horizon, sharply marked in a typical podsol, and in two distinct layers in a humus–iron podsol—an upper one (Bh) in which humus substances are precipitated and a lower (Bfe) in which ferric salts accumulate. The Bh layer is commonly dark chocolate brown or nearly black, and the Bfe layer reddish brown. Either or both may be cemented into a hard layer, the so-called 'iron pan', 'moor pan' or 'hard pan', which may be thick and resistant enough to stop the downward growth of tree roots and prevent their penetration to the richer subsoil, confining the roots to the impoverished A horizon. In some soils (peaty gleyed podsols) the iron pan is a very distinct narrow cemented band below the eluviated horizon (Plate 14). Foresters have to pierce or break up the hard pan when planting on such a soil, or else the planted trees do not make good growth.

In a 'podsol climate' podsols may be formed from almost any parent material, though rarely on limestones or highly basic igneous or metamorphic rocks, which provide an alkaline buffer as the rock is dissolved by carbonated water, so that acidification of the surface soil is prevented. However, on flat limestone surfaces in a sufficiently cool damp climate raw acidic humus may be formed from lichens and mosses colonizing the surface, and in this heath plants may germinate and take root, but a considerable thickness of insoluble residue must accumulate before a soil which can be podsolized is formed. In the warmer climates of the south and east, with their lower rainfall, podsols are commonly formed only on coarse sands and gravels of low base status, but these are frequently very well developed and show typical Bh and Bfe horizons, whereas in the nonsandy soils of the podsol climate so-called 'iron podsols', in which the Bfe layer alone is formed, commonly occur.

It must be emphasized that well-developed Brown Earth and Podsol profiles are to be found only in undisturbed mature natural soils where the conditions are favourable. There are many factors which may prevent or destroy the typical stratification described. Apart from destruction caused

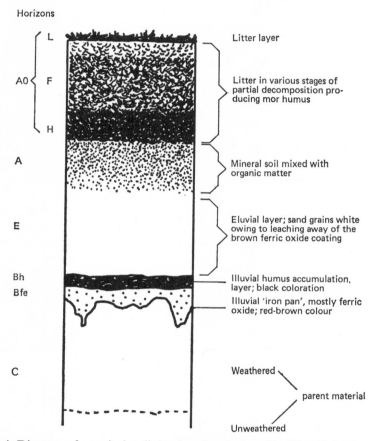

Fig. 15.4. Diagram of a typical well-developed podsol profile. The A0 horizon is subdivided into L (litter), F (fermentation) and H (humified) layers. The depth of such profiles is very variable, but is often about 1 metre. Podsols are common in areas of high rainfall and good drainage.

by ploughing, very many soils are immature, either because they are comparatively recently formed or because they exist under conditions, e.g. on steep slopes, where the soil can never become mature. Then again the surface layers of a mature profile may be destroyed by erosion, for example as a result of deforestation, and the lower horizons exposed. Such 'truncated podsols' in which the A horizon has been destroyed, exposing the B horizon on the surface, are not uncommon.

LOCAL SOIL TYPES

Besides the two main zonal (climatic) soils occurring in the British Isles and apart from the disturbed, eroded and immature soils just mentioned, there are various other soil types which are formed under definite local conditions. These soils are called *intrazonal* soils by pedologists and their formation is dominated by local factors such as topography and parent material.

Gley Soils. Soils which are seasonally or almost permanently waterlogged are, at least during periods of waterlogging, poorly aerated and under such conditions soil micro-organisms cause the chemical reduction of certain elements, particularly iron and manganese. Thus the manganic form of manganese changes to the manganous form and similarly the ferric form of iron changes to the ferrous form. When waterlogging ceases, air penetrates into the soil and causes the oxidation of the reduced elements. With such changes in the degree of waterlogging the soil shows signs of *gleying*, that is an alternation between oxidized and reduced conditions which are usually visible as colour mottlings in the soil profile. The anaerobic parts of the soil appear grey in colour (ferrous iron present) and the oxidized parts are rust-brown coloured (ferric iron). Often in gley soils small purplish-black concretions of manganese occur which are called manganese nodules and have the appearance of lead shot. Where gleying is limited to the lower parts of the soil profile the soil may be of the Brown Earth or Podsol type. Heavy clay soils in which drainage is impeded by the impervious nature of the soil to such an extent as to cause waterlogging are often referred to as surface-water gleys.

Peat Soils. When soil is constantly waterlogged and the water is relatively stagnant, oxygen is permanently deficient, and, in the absence of silting, organic material constantly accumulates and never disintegrates completely, thus forming a pure coherent organic soil or *peat*, often of great thickness. If the water has drained from calcareous or other basic rocks typical *fen peat* is produced, rich in bases, especially calcium, with a pH value between 7 and 8. This is formed in shallow 'fen basins' of which the largest British example, now almost all drained and cultivated, is found in the East Anglian fenland between Cambridge and the Wash. Undrained fen bears characteristic fen vegetation dominated by various reeds, rushes and sedges. If the plant debris accumulates so as to rise above the influence of the basic ground water, acid-loving calcifuge plants, notably bog moss (*Sphagnum* spp.), colonize the fen and form peat which is no longer saturated with calcium. This may serve as starting point of the formation of a *raised bog*, in which a kind of peat very different from fen peat, very poor in bases, acidic in reaction and inhabited exclusively by calcifuge species (see Chapter 16), is produced. This acidic *bog peat* is also formed in local depressions of heaths or moors, at the bottoms of moorland valleys where drainage is impeded (*valley bog*), and over wide tracts of flat low-lying country and on mountain plateaux where drainage is poor, rainfall of long duration, and the air almost constantly damp, so that the precipitation/evaporation ratio is persistently high (*blanket bog*). Blanket bog peat might indeed almost be reckoned as a climatic soil type since it depends on a decidedly wet climate; but on well-drained areas in the same regions podsols bearing heath, scrub, or even woodland are developed—absence of free percolation and of surface drainage ('run-off') being the other essential factor in the formation of blanket bog. If there is free downward drainage through the mineral substratum lying below the bog peat, podsolization will occur in it and A and B horizons develop. In such cases the peat may be regarded as an enormously exaggerated A0 horizon.

Rendzinas. Limestone soils form a class apart, as we have already seen (p. 170), owing to the overwhelming effect of the calcium carbonate of which the rock is mainly composed. On relatively pure limestones such as the chalk, they are characteristically shallow and dry. Below the dense dark surface humus, a young chalk soil is white or grey and passes down at once into the weathered limestone, the whole profile (Fig. 15.5), including the organic layer, being saturated with calcium carbonate. The pH value of a chalk soil is usually between 7 and 8. The unlimited supply of lime below acts as a strong alkaline buffer, lime being constantly brought to the surface by the upward movement set up when evaporation is strong, and by soil animals. In such a soil, called a *rendzina*, there is little or no leaching, and the humus is almost immobile, at the opposite extreme from the very mobile humus of a podsol. Consequently no A or B horizon is formed and the percolating rain-water, saturated with lime, drains away through the fissures of the rock. Such a soil is thus 'permanently immature'. It is added to from below by the gradual solution of the calcium carbonate leaving behind the insoluble mineral particles, of which even in the purest limestones a certain proportion is present.

In course of time a mineral soil thus accumulates on flat ground where it cannot be carried away by 'run-off', and more rapidly of course on impure limestones which contain a considerable proportion of insoluble mineral particles—sand, silt or clay. Ultimately, therefore, a substantial depth of mineral soil may accumulate over limestone in such situations, and leaching begins when the mineral soil reaches such depth that the surface layers are no longer affected by the underlying limestone. The effect of

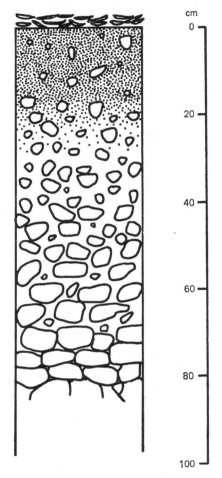

Fig. 15.5. Rendzina profile. The section is from beechwood on chalk escarpment. The surface layer is almost black with stable humus, and contains scattered lumps of chalk. Below, the lumps of weathered chalk are more numerous and closer together, with fine soil between, merging below with the parent rock. (Adapted from A. S. Watt, *J. Ecol.* 1934*a*, courtesy of British Ecological Society.)

surface leaching may be seen in some of the so-called chalk and limestone heaths developed on level expanses, as on the flat summits of chalk downs and on limestone plateaux. These show a curious mixture of species, heath plants growing in the surface humus while calcicolous species (see Chapter 16) are rooted in the calcareous soil below.

Besides these 'chalk heaths' extensive areas of the chalk plateaux in the south of England are covered with genuine heaths dominated by *Calluna vulgaris* and with no calcicolous plants. These are situated on loams or sands overlying the chalk, sometimes perhaps derived by continued leaching of an original chalk soil, but often composed of or including remains of non-calcareous Tertiary or Quaternary deposits which at one time overlay the chalk. An example of such a soil type is clay-with-flints, in which the soil matrix is derived from overlying beds and has become mixed with flints originally embedded in the chalk by the action of frost disturbance (cryoturbation) in Pleistocene times.

On the limestone soils of northern regions, where the climate is cold and wet, acidic humus often accumulates in the turf of the grassland so that the surface layer becomes acidic even close above limestone rock. In such grassland there are commonly a number of calcifuge plants which do not occur in chalk grassland.

ALLUVIAL SOILS

These soils belong to a third order known as *azonal* soils. The soils of this order lack a characteristic soil profile either because of immaturity or because conditions are not conducive to profile development.

Alluvial soils occupy flat ground with a fluctuating water table never very far below the soil surface. Typically this ground coincides with the alluvium of river valleys and is grassland (meadow) traditionally cut for hay, but now often used as permanent pasture. This riverside alluvium is subject to winter flooding which brings fresh silt to its surface and maintains its fertility. The meadow soils are in fact often rich in bases and very fertile. Flooding of the alluvial grasslands is now usually regulated or prevented by dredging of the river beds, by embankment, by the cutting of drainage ditches, and by regulation of the water flow through the sluices, but the ground water does not lie at any great depth. With a persistently high water table nearly reaching the surface and causing waterlogging, marsh plants tend to increase in abundance, but the true meadow soils are not deficient in oxygen owing to the constant rise and fall of the water. Impeded drainage and relatively high but fluctuating water-level are the characteristic features of alluvial soils. The abundant humus, largely derived from the numerous fine roots of grasses, remains near the surface and the underlying mineral soil is grey or blue-grey in colour. If this kind of soil is thoroughly drained so as to lower the water table considerably, leaching of the surface layer begins and the soil tends to develop the Brown Earth profile.

PRACTICAL EXAMINATION OF SOILS

In an examination of a soil in any area of natural or semi-natural vegetation the main concern of the plant ecologist is to determine those factors about the soil which may have direct bearing on the type of vegetation growing there. The following points indicate some of the more important soil characters which are of ecological importance and which should feature in any ecological investigation of the soil.

1. *General Notes.* A record should be made of the vegetation, geology, altitude, climate (especially rainfall) and topography (the surface relief including both slope and aspect) of the immediate area under investigation.

Field Examination

2. *Soil Profile.* A small pit should be dug in an apparently undisturbed place, from the surface down to the subsoil, so as to expose the whole soil profile. One side of the pit should be cut clean and vertical with a sharp spade or knife so that the stratification is not obscured or smeared. A sketch of the profile drawn to scale should now be made, with full notes of the general appearance, colour and texture of the various horizons. The colour of the soil may be expressed by use of the Munsell colour notation (Munsell 1954), the colour (in terms of hue, value or lightness, and chroma or strength) being matched against the colours given in the Munsell soil colour chart. The positions of the underground parts—roots and rhizomes —of the different plants growing on the surface should be indicated. The preliminary examination of the profile cannot be too carefully carried out, for it provides the primary data on which all subsequent work is based. It is wise to make sure that the spot chosen for digging the soil pit is truly representative of the soil of the area, and this can be done by making trial tests with a soil auger (or, where appropriate, a spade) in several different places. The soil should always be carefully replaced in all diggings, when the study has been finished.

The primary object of interest to the plant ecologist in the soil profile is the layer or layers in which the *feeding* roots of the plants are situated. Feeding roots can generally be identified by their appearance—their fineness, profuseness of branching and presence of root hairs. The general nature of the rooting layer will have been recorded in the profile sketch and some field tests can now be undertaken.

If detailed description of the properties of the soil horizons are needed, the scheme given for recognition and description of horizons in the *Soil Survey Field Handbook* (1960) (see Appendix) may be followed. Conventional symbols are used for reference to the various horizons, and an indication of the more important ones is given in Table 15.1.

3. *Soil Texture.* A great deal of information about soil texture can be obtained from the feel (between finger and thumb) of a moist soil; also by means of a simple key, such as that used at Field Studies Centres, textural class may be determined. For example, if a soil feels gritty it is a sand or sandy loam (the latter, but not the former, holds together in a ball in the

hand); if it is sticky or silky or both it is a clay (smooth shiny surface) or silt loam; if it is sticky and gritty it is a sandy silt loam or sandy clay loam; and if it is not gritty, sticky or silky it is a loam (cf. Fig. 15.1). In general, grittiness denotes sand, silkiness denotes silt and stickiness denotes clay as the dominant texture.

More quantitative estimations of soil texture can be made on samples of soil removed to the laboratory where the proportions of particles of different sizes in the different horizons can be determined (see p. 167).

Table 15.1. Soil profiles: notation for horizons

Organic and organo-mineral surface horizons

A0	Organic horizon above mineral soil; may be sub-divided into:

	L	Plant litter, not decomposed
	F	Litter partly decomposed (fermentation) and comminuted
	H	Humus well-decomposed, low in mineral matter

A	Mainly mineral matter with humus admixture
Ap	Ploughed layer

Sub-surface horizons

E	Eluvial horizon, depleted of clay and sesquioxides
Ea	Bleached, pale horizon, depleted of iron and aluminium
B	Altered horizon distinct in structure or colour, or enriched in humus, iron and aluminium or clay; may be sub-divided into:

	Bh	Dark horizon, enriched in humus
	Bfe	Red-brown horizon, enriched in iron (and aluminium)

C	Little-altered parent material

---

4. *Calcium Carbonate.* Pouring a few drops of dilute hydrochloric acid (e.g. 20 per cent) on the soil to be tested gives useful information as to whether or not the soil can be described as calcareous (see p. 83). The degree of effervescence which results is directly proportional to the amount of calcium carbonate present. Numerous tests on different layers and on the same layer in different places can be made in a short time and thus the distribution of any free calcium carbonate present rapidly determined. If there are comparatively few large particles in the sample the bubbles of gas will be local but vigorously evolved. If there is a much greater number of small particles—of immensely greater effect in neutralizing acids because of the greater aggregate surface—the evolution of gas will be less vigorous but more general. Considerable local differences may often be observed in carbonate distribution, both from place to place and especially with depth from the surface, owing to progressive leaching of the surface layers. These

differences can sometimes be correlated with the rooting positions or layers of the different species.

5. *Degree of Acidity*. A brief description of the meaning and significance of pH value has already been given (see p. 165). Dependent on the type of equipment available, pH determinations may be carried out in the field or in the laboratory.

A good indication of soil acidity can be obtained by flooding a small sample of soil with different indicators (BDH Universal Indicator is a mixture of such indicators). Special soil testing kits such as BDH Soil Test Outfit use glass tubes in which samples of soil are shaken with distilled water and a few drops of indicator solution; barium sulphate may be added to encourage flocculation and settling of the clay particles. When the particles have settled, the colour of the clear solution is compared with a standard colour chart. More accurate determinations of soil pH values can be made by means of a glass electrode and a portable pH meter. A number of factors affect the pH reading obtained, especially the salt content of the soil, the proportion of soil to water in the sample, the temperature and the carbon dioxide concentration; consequently pH measurements on water suspensions of soil samples taken in spring or summer may give different results from similar measurements made in winter. It should also be remembered that the soil is a very heterogeneous system and extreme accuracy in single determinations is not of great value. The method of preparation of the sample and of determination must be standardized for comparative purposes and it is preferable and more meaningful to make a series of determinations on a soil so that some indication of the degree of variation in pH values for that soil is obtained.

6. *Deficiency of Cations of Bases*. Comber's reagent (potassium thiocyanate in alcoholic solution) gives a rapid test of base deficiency in an iron-containing soil. When a sample of soil is shaken with this reagent, ferric ions come into solution as a result of ion exchange and red ferric thiocyanate is formed if bases are deficient. If the soil is not deficient in bases, ion exchange brings calcium and magnesium ions into solution rather than ferric ions. The intensity of the red colour provides a good indication of the degree of base deficiency.

It should be noted that, because Comber's reagent actually measures the amount of ferric iron brought into solution, it is valueless where iron is nearly or quite absent, as in some peaty or humus soils.

A modification of Comber's test can be used as an indicator of anaerobic soil conditions. Under anaerobic conditions microbiological activity changes ferric iron into the ferrous form. Ferrous iron does not give a colour reaction with potassium thiocyanate but if ferrous iron is present in the soil solution it can be converted to the ferric state by adding one drop of hydrogen peroxide—a strong oxidizing agent—to the solution. If the red colour obtained before addition of the peroxide is now intensified then ferrous iron, and therefore reducing conditions, are present in the soil.

Simple Laboratory Determinations

7. *Water Content and Water-holding Capacity.* The amount of water present in the soil should be determined in two stages. Ten grams of fresh soil are weighed out, spread out on a sheet of paper in a dry room, and then weighed again after an appropriate interval. When the soil no longer loses weight it is said to be 'air-dry', and the water lost can be expressed as a percentage of the weight of the fresh soil. This of course will vary according to recent rainfall or proximity of ground water and it is important that the samples are carefully collected and transported to the laboratory in a container with a tightly fitting lid or in a polythene bag, and also that they are clearly marked for identification. The air-dry soil should then be heated in an oven kept at 100° C. and the further loss in weight recorded. This can be expressed as a percentage of the air-dry soil which is fairly constant for a particular soil or the total loss in weight may be expressed as a percentage of the oven-dry soil weight.

Water-holding capacity may be determined by carefully saturating a sample of soil in a sieve, allowing it to drain freely and then drying the sample in the oven at 100° C. Water-holding capacity, as loss in weight after drying, may be expressed as a percentage of the oven-dry weight of soil.

8. *Loss on Ignition.* The loss in weight when a soil (previously dried at 110° C. rather than 100° C. in this instance) is heated to redness in a crucible represents the organic matter content together with the 'water of constitution' of the clay present, and the carbon dioxide lost by the decomposition of calcium carbonate. This method is therefore unsatisfactory for calcareous soils unless the calcium carbonate content is determined separately and allowance made for the carbon dioxide set free from the carbonate during burning. If a controlled temperature laboratory furnace is available it is preferable that ignition should be maintained at between 400° C. and 450° C. over a period of from 7 to 8 hours. Loss on ignition is not a specific method for determining the organic matter content of soils but with certain precautions it gives useful comparative results for non-calcareous soils.

Other Determinations

Other soil properties are often measured but these require more specialist knowledge and equipment and discussion of the methods are beyond the scope of this book, Such determinations are of exchangeable bases, e.g. calcium, magnesium and potassium; of trace-element content, e.g. man-

ganese, boron, zinc; of nitrogen (often determined by the Kjeldahl method
for total nitrogen); and of phosphorus content. Measurement of the composition of the soil atmosphere is not a simple matter and special equipment
is needed to produce meaningful results. For further information on
methods of soil analysis reference may be made to more specialized texts,
such as those by Jackson (1962) and Hesse (1971).

Chapter 16

Edaphic Factors: Mineral Nutrition

Studies on the distribution of plants in the British Isles indicate clearly the substantial influence which soil may have on the occurrence and growth of plants and also reflect the wide range of types of soil which are represented. As already indicated in Chapter 15, soils may differ markedly in physical features, several of which have a considerable effect on the growth of plants, but also soils vary very markedly in their chemical composition, which is of major importance with respect to plant distribution.

Firstly, different soils may contain and provide in available form different levels of essential elements. Within limits, the higher the levels the more chemically fertile is the soil. Levels required for the optimum growth of plants are rarely found in the field and frequently the ecologist is dealing with marginal land, so called because it is of little use agriculturally and is deficient in one or more of the essential elements. Secondly, although plants absorb elements selectively, the balance, or the difference in proportions, of essential elements in the soil, can have a strong influence on the distribution of plants. A prime example concerning calcium levels is discussed below. Thirdly, an excess of certain polyvalent metal ions, some of which are essential for growth in concentrations of less than 1 p.p.m. (e.g. manganese, iron, zinc and copper) and some which are not (e.g. aluminium and lead), will prevent the establishment of many species of plants.

The ecologist who studies mineral nutrition is therefore concerned largely with the *tolerance* of plants to nutrient stress due to deficiencies or to chemical toxicities.

The simplest and most revealing single measurement which can be made with respect to the chemical characters of a soil is of pH. The optimal figure for plant growth, from an agricultural point of view, is around pH 6·0–6·5. Such a pH value usually denotes a suitable balance of essential elements; polyvalent ions are present only in trace amounts, and the establishment of crop plants and ruderals (plants of waste ground), both of which usually have high relative growth rates, is favoured. Low pH values generally imply deficiencies of essential nutrients and an excess of potentially toxic cations, such as of aluminium, and, for some plants, ferrous iron. High pH values imply calcareous or saline soils (above pH 8·0), the former with an excess of calcium, the latter with an excess of sodium ions, and both frequently have deficiencies of certain trace elements. Plants which can survive such extremes are usually inherently slow growing. It is of interest that present experimental evidence does not indicate a critical role of either hydrogen or hydroxyl ions specifically. Factors which are believed important are considered with further reference to calcareous and non-calcareous soils.

CALCICOLES AND CALCIFUGES

It is well known to all botanists that the flora and vegetation of highly calcareous soils, such for instance as the shallow (rendzina) soils of the chalk downs, are very different from those of markedly acidic, base-deficient soils like those of a typical sandy heath or of a peat-covered moor. Many of the species found on calcareous soils rarely or never occur on highly acidic soils, and vice versa. These two categories of species have long been known respectively as *calcicoles* and *calcifuges*. There are also a number of species which have been found to survive on examples of both of these extreme types of soil. They appear to be indifferent to a wide range of lime content and its associated soil reaction.

For a long time there was a difference of opinion as to whether the characteristic species of limestone soils inhabited them because of the distinctive chemical conditions or whether they depended rather on the distinctive physical conditions of limestone soils.

Evidence for both situations is now available with chemical factors seen as the initial source of failure and physical conditions such as frost and droughting the ultimate ones. Hope-Simpson (1938) showed that many characteristic chalk species also occurred in quantity on a certain plot of Lower Greensand soil which in the past (probably well over half a century before) had been so heavily limed that its average content of calcium carbonate when the observations were made was distinctly high (4·5 per cent) though not nearly so high as that of a typical chalk grassland soil, which often exceeds 30 per cent. Since the *physical* nature of this lime-containing Greensand soil was virtually the same as that of a comparable plot of the normal Greensand soil which contained *no* detectable calcium carbonate (though minute quantities of calcium ions were present) it is difficult to avoid the conclusion that the 'chalk plants' present on the lime-containing Greensand are genuine *chemically determined calcicoles*. Among these were such well-known and abundant chalk grassland species as the hairy hawkbit (*Leontodon hispidus*), salad burnet (*Poterium sanguisorba*), upright brome (*Zerna erecta*), and a number of others.

Besides the 'chemically determined calcicoles' on the plot of lime-containing Lower Greensand soil there were a number of species, constant in chalk grassland, but also occurring on the normal Lower Greensand soil free from lime. Among these were the two fine-leaved fescues (the sheep's fescue, *Festuca ovina*, and the red or creeping fescue *F. rubra*), and such plants as ladies' bedstraw (*Galium verum*) and ribwort plantain (*Plantago lanceolata*). The fine-leaved fescues are the commonest dominants of pastured chalk grassland, but they also occur on other dry soils with little or no calcium carbonate, the sheep's fescue being notably co-dominant with the common bent-grass (*Agrostis tenuis*)—which is on the whole a calcifuge—over very wide areas of rough pasture on the siliceous rocks of the west and north, and on grazed sandy commons in the south. Ladies' bedstraw is far from being confined to chalk grassland, and ribwort plantain is an exceedingly common plant found on a wide variety of soils. These species are certainly not chemically determined calcicoles, though they are abundant on soils containing an excess of calcium carbonate. It is of considerable interest that for species such as *Festuca ovina*, which occur in a range of grassland communities with soil pH values of from less than 4 to more than 7, there is now evidence that physiological races or *edaphic ecotypes*, differing in their response to calcium, are involved (Snaydon and Bradshaw 1961). Similarly the occurrence of races of *Agrostis tenuis* having distinct tolerances of different heavy metal toxicities has been demonstrated by Gregory and Bradshaw (1965).

When we turn to the typical *calcifuges*, we find a strong intolerance of lime-containing soil water. Of characteristic species belonging to this category the common ling (*Calluna vulgaris*), the wavy hair grass (*Deschampsia flexuosa*) and the heath bedstraw (*Galium saxatile*) occurred on the acidic Lower Greensand but not on the calcareous sand, and they are never found on genuine chalk soils. Most species of bog moss (*Sphagnum*) are also highly intolerant of lime-containing water. Many species of the heath family besides the common ling never occur on calcareous soils, for example the purple bell heather (*Erica cinerea*) and the allied bilberry (*Vaccinium myrtillus*), and the same is true of many bog-plants such as the cottongrass (*Eriophorum*).

What is it that determines the limitation of some species to soils containing abundant calcium carbonate and others to soils with undetectable quantities of calcium carbonate? Experimental studies (Rorison 1960) indicate that neither pH *per se* nor the concentration of calcium ions is important alone—although it is known that a calcicole such as the marjoram, *Origanum vulgare*, has a higher requirement of calcium for *optimum* growth than a calcifuge such as the heath rush, *Juncus squarrosus* (Jefferies and Willis 1964). It is rather the combination of chemical factors associated with soil pH which often has a predominant influence on plant distribution.

Calcicole species such as the small scabious, *Scabiosa columbaria*, fail on well-aerated acidic mineral soils for a number of reasons. If the pH of the soil solution is sufficiently low ($<5\cdot0$) for polyvalent ions such as aluminium and manganese to be in solution then they are liable to be toxic. Ferric iron is virtually insoluble at pH values greater than $3\cdot5$ and is therefore unlikely to be toxic in well-aerated soil but in acidic waterlogged conditions the more soluble ferrous iron may be a major toxin to calcicoles. Uptake of phosphorus is inhibited in the presence of aluminium. Levels of calcium are likely to be sub-optimal and more recently it has been shown that nitrogen in the ammonium form can also be toxic under acidic conditions.

Seedlings of calcicoles growing under these conditions are typically stunted with few, discoloured leaves and with only few and peg-like lateral projections characterizing very under-developed root systems. They may survive in this state for long periods but ultimately are killed by desiccation or soil movement resulting from the action of rain or frost (solifluction). Laboratory experiments (Rorison 1967) have shown that even the germination of strict calcicole species can be affected by acidic soil conditions.

Conversely, the most highly calcareous soil does not affect the *germination* of naturally occurring species, even of such a strict calcifuge as wavy hair grass (*Deschampsia flexuosa*). *Seedling survival*, however, is affected to a degree which can be related to the level of calcium carbonate in the soil.

Root systems of calcifuge species have a deformed appearance very similar to that of calcicoles in acidic soils (Hutchinson 1967). In this case stunting is known to be related to inhibition of iron uptake by bicarbonate ions (Lee & Woolhouse 1969). Leaves of susceptible species growing on calcareous soils typically turn yellow—a condition often termed *lime-induced chlorosis*. Its exact cause remains uncertain. It occurs, in a transitory form, even in calcicole species at certain times of the year (Grime and Hutchinson 1967) and characteristically is shown in young, actively growing leaves and in apices at the extremities of plants, such as the unrooted runners of the mouse-ear hawkweed (*Hieracium pilosella*), suggesting that the degree of mobility of iron, which is necessary for the production of chlorophyll, may be concerned.

Contributory factors to the failure of calcifuges on calcareous soils may also include an excess of calcium which affects uptake of other cations, particularly potassium. An excess of nitrogen in the nitrate form has been shown experimentally to be deleterious to the growth of some strict calcifuges at pH values greater than 7·0.

The importance of the levels of mineral nutrients in soils in influencing the occurrence, growth and success of plants can also be seen when mineral salts are added to the sites where some nutrients are in short supply. The enhanced levels of minerals, which make good certain deficiencies in the soil, often lead not only to much-increased growth but also to pronounced changes in the floristic composition of the vegetation, some plants becoming strongly dominant and others, often smaller and more slow growing, being eliminated. In the coastal dunes of Braunton Burrows, North Devon, for example, the addition of mineral nutrients which increased the levels of nitrogen, phosphorus and potassium, all elements in short supply in the sandy soil, enhanced the yield of the vegetation as much as sevenfold. It also resulted in the strong growth of grasses such as creeping fescue (*Festuca rubra*) and creeping bent (*Agrostis stolonifera*) but the reduction of many low-growing rosette species and bryophytes (Willis 1963). When deficiencies were made good except that of phosphorus, sedges rather than grasses increased substantially, indicating a difference in response to levels of mineral nutrients by different species.

Evidence of the importance and the role of mineral nutrition in the distribution of plants is continually growing but we are still some way from a thorough understanding of all the individual factors and their inter-relationships.

RANGE OF TOLERANCE AND COMPETITION

Soil conditions are complex; local climate and water supply exert important effects, different factors interacting with one another (p. 143). Different species differ greatly in their requirements from the soil, and show various degrees of tolerance of soil conditions. Clearly the actual sites in which the individuals of particular species occur, i.e. which the seeds can reach and in which they can germinate and the seedlings establish, may vary considerably in their suitability for that species. Some plants have a narrow, some a wide range of tolerance of various soil conditions. Take for example the factor of acidity and alkalinity, i.e. of soil reaction. It has been shown by careful culture experiments with a number of species that each has a definite range of pH value within which it can grow, and a narrower (optimum) range within which it grows luxuriantly. Within the wider range, but outside the narrower, the plant is evidently more or less limited in its growth and metabolic functions. The factor of *competition* with other species will play a decisive role in this situation. A plant may be able to succeed in a soil of a certain soil reaction when it is growing alone, but not if this pH value, with the rest of the factors present, enables surrounding plants to grow more vigorously so that they overshadow and tend to smother it.

It is now realized that strict calcicoles and calcifuges are better characterized by their tolerance of extreme environmental conditions—such as high levels of polyvalent ions at low pH—rather than by any particular requirement for optimum growth. Their optimum relative growth rates are in fact low compared with many more widespread species and they are not found in soils of intermediate pH—fertile or infertile—because they lack the competitive ability to survive even though they grow best there.

In a neglected garden bed many of the garden plants, which grow very well when isolated, are soon smothered by weeds if these are not removed. The weeds have a much wider range of tolerance in many respects and many garden plants are growing under conditions which are not optimal for them, though they succeed quite well in the absence of competition.

One of the first experiments illustrating the relationship between soil type and plant competition was reported by Tansley (1917). The bedstraws *Galium saxatile* and *G. pumilum* (*sylvestre*) were grown separately and together on an acidic and on a calcareous soil. In nature *G. saxatile* flourishes on, and is confined to, non-calcareous soils whereas *G. pumilum* tolerates large amounts of calcium carbonate in the soil. When grown alone, both species survived on both soils, but when grown together, the introduced species was ousted because the species normally occurring on the soil tolerated the conditions much better and its growth rate was unimpaired.

Similar experiments have helped to determine the causes of the distribution of other species and experimenters are now proceeding to the investigation of the physiological causes of the actual effect of the soil on the species.

As with all ecological problems, the complexity of situations in the field has to be reconciled with the necessarily simplified conditions of laboratory experiments designed to solve specific problems, and the relationship between mineral nutrient factors and the climatic and biotic influences acting on the soil environment must never be overlooked.

Chapter 17

Biotic Factors: Nature and Interaction of Ecological Factors: Habitat Changes

BIOTIC FACTORS

The biotic factors of the habitat are those which depend directly on the action of living organisms on the vegetation (Greek βιωτικός, pertaining to life, βίος). It is obvious at once that here we are involved in a logical difficulty, for the plants included in any community may, as we have seen, have a profound effect upon one another, just as the individuals of a human community have such a mutual effect. In order to get a practicable working definition we need to exclude the mutual effects of members of the community, and apply the concept of biotic factors of the habitat, i.e. of the community habitat, only to the action of organisms which are not regarded as part of the community. But then we are confronted with the question of which organisms are to be regarded as part of the community and which are not. Are these the soil bacteria; earthworms and other soil animals; parasitic fungi; the snails and insects that live on the plants or in the soil; the birds that live in the trees and may play an important part either in destroying or in disturbing fruits or seeds, or in doing both? Directly we put such questions we begin to realize how closely interwoven is the web of nature, and how artificial our distinctions and classifications in reality are.

The most natural concept that has been suggested is that which regards the whole complex of organisms—both animals and plants—naturally living together as a sociological unit which has been called the *biome*, and whose life must be considered and studied as a whole. The biome will include not only soil algae, bacteria, and earthworms, not only insects and parasitic fungi, but rabbits, mice and other rodents. In the semi-natural pasture communities, which are maintained in the condition of pasture by grazing, we must include the sheep or cattle which are regularly pastured upon it, and which, as we have seen, are the chief factor keeping this plant community in equilibrium. Thus for practical purposes it is necessary to regard separately, and to study separately as a 'biotic factor', any collection or group of animals which have marked effects upon a plant community.

The effects of such animals can be studied in various ways. By acute observation alone the good field naturalist can learn much about them qualitatively, but seldom can say exactly how much influence they have, especially where a number of different animals are concerned together. To

195

obtain such knowledge it is necessary to make exact experiments, by excluding a given species or type of animal from a portion of the plant community.

Small vertebrates can be kept out of a small plot by enclosing it with a wire-netting fence of suitable mesh. One-inch (2·5 cm) netting will keep out rabbits, while netting of ¼-inch (0·6 cm) mesh is necessary to exclude mice, and the wire must be sunk well into the soil. The height of the netting above the ground-level should be 3 feet (90 cm) for rabbits, and at least 12 inches (30 cm) for mice and voles. Mice and voles, however, sometimes climb fences of this height, and 2 feet (60 cm) netting is desirable. They can be excluded by vertically placed sheet-iron plates, by use of polythene placed over the top edge of wire netting, a galvanized iron outwardly-deflected 'roof' on the netting, or by covering in the exclosure with ¼-inch (0·6 cm) netting, which also keeps out birds.

Demonstrations of the effect of sheep grazing on upland grassland may be seen in some National Nature Reserves, e.g. Cwm Idwal in North Wales. In these experimental plots some areas are fenced off (preventing grazing) at all times while others have movable fences so that grazing can be controlled and confined to specific times of the year. The most obvious responses to the complete exclusion of sheep are the increase in the height of the vegetation, the invasion of shrubby species such as heather (on acidic soils) and a decrease in the number of small low-growing herbs. The effect of sheep grazing on the vegetation of upland heaths in Scotland has already been noted (Chapter 12). In wooded areas, sheep grazing may affect the survival of tree seedlings, and also strongly influence the shrub layer and ground flora. Under severe grazing conditions natural regeneration may cease entirely, but with the exclusion of sheep from such areas tree seedlings may again successfully establish and there may be pronounced changes in other components of the vegetation (Plate 15).

Invertebrates are of course much more difficult to exclude than mammals, especially those, such as slugs and snails, which live in or wander through the soil instead of merely traversing its surface. Wire gauze can sometimes be used with advantage for short periods, but this cuts down illumination and prevents the entrance of seeds and fruits into the protected plot. The methods of preventing the access of invertebrates must be left to the ingenuity of workers according to the particular animals to be excluded and the particular conditions of the experiment.

Extensive field work by animal ecologists, especially the Oxford Bureau of Animal Population, has demonstrated the existence of great cyclical fluctuations in the populations of such small rodents as voles and mice and, since these animals may be very destructive, the marked increases and decreases in their numbers are naturally of great importance in influencing the vegetation.

Animals may affect vegetation in a number of ways: they may eat and damage it so as to cause, in an extreme case, the replacement of one community by a totally different one, and they may act as pollen or seed distributors; they may also, for example, alter the soil by manuring, or by loosening or compacting it. Grazing by virtually all herbivores is a selective process, the more palatable plants being eaten first. Some species, such as white clover (*Trifolium repens*) and bird's-foot trefoil (*Lotus corniculatus*), are known to have forms which are cyanogenic, i.e. contain a chemical compound which forms hydrocyanic acid when the stems or leaves are injured. Invertebrate animals (e.g. snails) and small vertebrates (e.g. voles) may avoid plants containing this compound in preference for the acyanogenic form.

There is good evidence that the regeneration of grasses and other useful grazing plants of the American cattle-ranges in some of the Western States is promoted by the trampling of cattle after the seeds are ripe. The view has been held that the presence of swine in English oakwoods and beechwoods may have helped their regeneration from seed. Although the pigs feed on the mast, it is possible that they favour regeneration by trampling some of it into the soil and thus enabling the seeds to germinate; they may also eat various animals such as mice and slugs which attack the seeds and seedlings. There is abundant evidence that the destruction of carnivorous birds (hawks, jays, etc.) and small animals (stoats, weasels, etc.) by gamekeepers handicaps or destroys the chances of tree regeneration in many woods, because mice and voles, and before myxomatosis, rabbits, which are part of their natural prey, multiply in their absence and destroy the seeds and seedlings of the trees, especially oak and beech. Thus an interference with the balance of the animal population may alter the relation between one part of it and the dominants of the plant community, and eventually result in the disappearance of the latter.

NATURE AND INTERACTION OF ECOLOGICAL FACTORS

In studying the different factors of the habitat and estimating their combined effect on the vegetation, certain principles must always be borne in mind.

The first principle is that the forces which can actually affect plants are limited in number and nature by the constitution of the plant itself—in the case of the ordinary rooted land plant by those features which are common to all such plants—and also by those peculiar to the species. Apart from 'gross' factors, such as damage or destruction by wind, frost, fires, grazing animals, insects, and parasitic fungi, when we speak of 'factors' of the habitat such as rainfall, water content of soil, or the kinds and amounts of mineral nutrients present, we must remember that they can be effective in only a limited number of specific ways. These are ultimately reducible to water, with its content of free ions capable of absorption by (or otherwise affecting) the roots, free oxygen, carbon dioxide, available light falling on the leaves, evaporating power of the air, and temperature. The term 'factor' is in fact used in ecology for any substance, force, or condition affecting the vegetation directly or indirectly in such a way as to differentiate it from other vegetation, and the so-called 'factors' have always to be ultimately interpreted in terms of the mechanical, physical and chemical processes directly concerned in the life of the plant.

In regard to the water turnover of the plant, a reduced evaporating power of the air means that the roots will make a smaller demand for water on the soil, so that in a region with constantly humid air the same species can grow on soil which has a lower water supply. Direct evaporation from the soil itself in a humid climate will also, of course, be much less, so that the soil will maintain an adequate water supply with a much lower rainfall than would be necessary in a hot dry climate. Thus a sandy soil in the west of England, where both rainfall and air humidity are high, will support a more moisture-loving vegetation than the same soil in the eastern counties. Again, in eastern England most soils, because of the moderately high air humidity, will, with a rainfall of 20 inches (500 mm), support a more luxuriant vegetation than in a continental climate, where similar soils with the same rainfall might produce only a dry grassland. Similar relations of air and soil moisture may be seen on a smaller scale wherever the physiographical features produce sharp differences of 'local climate', as in steep ravines and on exposed ridges.

Replacement of One Ecological Factor by Another
Such instances as those just mentioned illustrate what has been called the replacement (or compensation) of one factor by another. Further, many of the same species occur on the southern exposures of hills with dry shallow soils in western and western-central Europe as on various different soils and exposures in the drier Mediterranean climate. Here the local physiographic and edaphic factors replace the general climatic factors of the Mediterranean region. Inversely, steep northern exposures and deep ravines in the Mediterranean region often bear a distinctly northern vegetation. The stemless thistle, *Cirsium acaulon*, occurs on calcareous soils and is common in the south-east of England (Fig. 17.1). Its distribution seems primarily determined by climate. In Derbyshire and the Yorkshire Wolds at the northern limit of its distribution in Britain this species is found predominantly on south- to south-west-facing slopes whereas in Kent and Sussex it occurs on slopes of all aspects. As indicated above, it is well known that south-facing slopes (in the northern hemisphere) have temperature conditions which match level sites of much more southerly latitudes. Pigott (1968) has shown that *Cirsium acaulon* is dependent on high summer temperatures for the production of ripe fruit and the species is limited in its distribution in Britain to areas where the summer daily

maximum temperatures exceed 21° C. and 20° C. during July and August.
Further north and west, a combination of lower maximum temperatures
and higher rainfall prevents the occurrence of sufficiently high temperatures inside the flower head for the successful maturation of the fruits. An
interesting point is that, at the more northerly localities of the stemless
thistle, in those years when ripe fruit are produced they are restricted to
the south-facing edge of the capitulum where the temperatures are greatest.

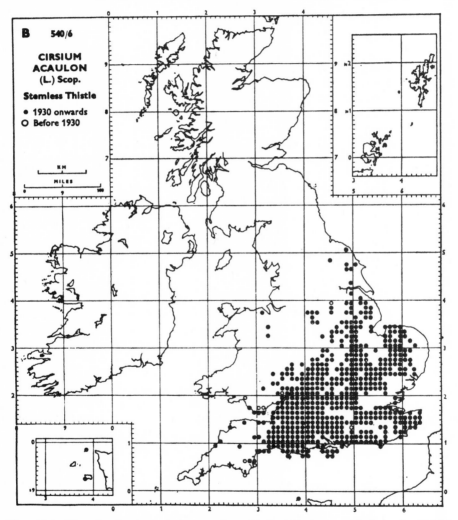

Fig. 17.1. The distribution of the stemless thistle (*Cirsium acaulon*) in the British
Isles. Each dot represents at least one occurrence in a 10-km square of the National
Grid. (By courtesy of the Botanical Society of the British Isles and Thomas Nelson &
Sons Ltd from the *Atlas of the British Flora*.)

Limiting Factors

The second principle is what has been called the 'law of limiting factors' which is of great importance in ecology. When the co-operation of two or more 'factors', e.g. definite quantities of different substances, temperatures within certain limits, or a particular intensity of illumination, are required for the maintenance of any process, then, if one of the 'factors' is absent, the process stops whether the others are present or not. Similarly, if the rapidity of a process varies with the quantity or degree of several different 'factors' acting together, the reduction of any one of them will reduce the rate of the process, even though the others are present in sufficient quantity or degree, or in excess.

For example the necessary mineral nutrients may be present in the soil, and temperature and light may be favourable; but if water is absent the plant cannot grow, and if minimal amounts are present it can grow only very slowly, as in a desert. Again, if the temperature remains below freezing-point, plants cannot grow, though all the other factors may be favourable. And as the temperature rises above freezing-point growth will steadily increase up to a certain point if the other factors are maintained at a suitable level. It may then, for instance, be checked by the water factor, the water supply in the soil being adequate to maintain the slow growth which can take place at 5° C., but inadequate for the more rapid growth which would be made at 15° C. or 25° C. if sufficient water were available. The rise in temperature will also tend to decrease the water supply in the soil by increased evaporation, both directly through the soil and also through the plants, and this effect (increase of transpiration) may render the plant unable to cover its water loss by absorption from the soil. A similar situation exists with respect to the other necessary factors, carbon dioxide, light, and essential mineral nutrients. The factor which is present in so low a quantity or degree that it limits growth or some other life process is called the *limiting factor*.

HABITAT CHANGES

A further point about the habitat should always be kept in mind—the fact that it may *change* progressively, both along with the vegetation and more or less independently also. This point has already been considered in Chapter 4, where we saw that succession or development of vegetation from a starting-point on bare ground was normally accompanied by a gradual increase of humus in the soil, and that this increase in its turn progressively fitted the ground for different kinds of vegetation until a climax was reached. On all the more favourable soils this climax tends to be determined by climate (climatic climax), and to represent the highest (most complex) type of vegetation that can exist in the general climatic conditions. We have seen in Chapter 15 that organic matter is of great importance in 'improving' the soil, provided there is a quick 'turnover', and so helping to give the most favourable conditions for vegetation. If these favourable conditions are maintained indefinitely the climatic climax will be correspondingly maintained; the ecosystem will be mature.

Other factors may, however, come into play. In a cool humid climate, such as that on and near the western coasts of the British Isles, the organic matter does not disintegrate as rapidly as it is formed, and thus tends to accumulate as mor (see p. 172). In the wetter places this results in the formation of peat, and bogs and moors occur instead of forest. This is most marked over soils poor in lime and in mineral bases generally, because the presence of cations of bases tends to promote the growth and activity of soil organisms which disintegrate the organic matter and provide suitable conditions for plants other than those typical of bog and moor. But under cool and extremely moist conditions, especially if the soil is badly aerated, raw humus and peat may form even over limestone itself. Here we have good examples of the replacement of factors already described. A cool moist climate, local excess of soil water, poverty in mineral salts, and poor aeration, are all factors tending in one direction, and may to a considerable extent replace one another.

Leaching

Another factor tending to alter the habitat in the same direction, and having a considerable effect in the British Isles, is the leaching or washing out of soluble mineral salts (see Fig. 15.2, p. 176), already mentioned in Chapter 15. Different salts are soluble to very different extents. Sodium chloride and other salts left in maritime soils wash out the most readily, and so alter the habitat presented by these soils when they are no longer supplied with fresh salt from the sea. Of the mineral salts present in ordinary soils, calcium carbonate is most readily leached out by water containing much carbon dioxide, while magnesium and potassium salts dissolve much less quickly. The impoverishment of the surface layers of soil in calcium carbonate and to a less degree, in other bases, will lead to 'acid conditions' and to the establishment of acid-tolerant plants, and may gradually change the whole character of the vegetation. It is believed that much of our semi-natural vegetation has been, and is being, steadily altered in this way. In wet regions leaching will clearly assist peat formation, in drier ones it will slow down the turnover of organic matter and thus tend to the accumulation of mor, for instance on a forest floor; and this may help, along with other factors, to prevent the regeneration of the forest, or lead to the replacement of one species of dominant tree by another.

The nature of the plant litter which is deposited on the soil surface also influences the formation of mor. The leaves of woodland trees, for instance, differ very considerably in their resistance to the process of decay and this is broadly correlated with their acidity. Conifer needles, particularly, are very resistant, so that the litter which they form tends to accumulate and decomposes very slowly. This fact has an important bearing on the large plantations of conifers because these trees tend to change the type of humus in the soil from mull to mor, increasing the acidity of the surface soil, promoting leaching and degradation of soils from the Brown Earth to the Podsol type. Most broad-leaved deciduous trees form litter which decays much more rapidly and the leaves contain more bases such as calcium, magnesium and potassium, so producing much less acidic organic matter in the soil. Beech litter, with a pH value of 6·6, is more resistant to decay than oak litter, and on a sandy soil, poor in bases, easily develops mor, though on a loam of high base status it forms a good mull. Plants such as ling (*Calluna vulgaris*) and gorse (*Ulex europaeus*) have been shown to produce litter which has the effect of acidifying the soil below the bush and, on initially less acidic soils, for example those supporting limestone heath communities, this may result in an overall change in species composition of the vegetation cover (Grubb, Green and Merrifield 1969).

Chapter 18

The Ecosystem: Nutrient Cycling, Energy Flow and Productivity

A wider concept than the community and the biome is to include with the biome all the physical and chemical factors of the biome's environment or habitat—those factors which we have considered under the headings of climate and soil—as parts of one physical system, which we may call an ecosystem, because it is based on the οἶκος or home of a particular biome. All the parts of such an ecosystem—organic and inorganic, biome and habitat—may be regarded as interacting factors which, in a mature ecosystem, are in approximate equilibrium: it is through their interactions that the whole system is maintained. Thus the ecosystem is viewed as a working system and the idea of a dynamic interrelationship between the physical, chemical and biological components has given considerable impetus to the study of ecology.

For the plant ecologist the concept of the ecosystem, an ecological unit of any rank, is valuable in giving a truly biological perspective. The influence of environmental factors on the ecosystem as a whole can be assessed and so the consequences of the alteration of the environment on the delicate balance in the ecosystem predicted. The ecosystem approach focuses attention on the interactions between the various organisms and the environment and the movement of all materials utilized by the biome can be followed in this broader context.

NUTRIENT CYCLING

Many substances, such as mineral nutrients, water, oxygen, carbon and nitrogen, are continuously recycled within the ecosystem. There is also an energy flow through the ecosystem, but this is a one-way and down-hill process. Some of the major features of nutrient cycling and energy flow are shown in Figs. 18.1, 18.2 and 18.3.

As already discussed (Chapter 15), mull humus derived from dead plant material is rapidly broken down, under suitable conditions, to give carbon dioxide, water and simple inorganic salts. This process, by which the reserves of those elements needed for healthy plant growth are continually replenished, is known as *mineralization*. Mineralization, however, is not the only source of inorganic nutrients, for, as indicated earlier, weathering of rock and rock minerals in the soil continually adds their constituent elements in simple ionic form. In the comparatively young or shallow

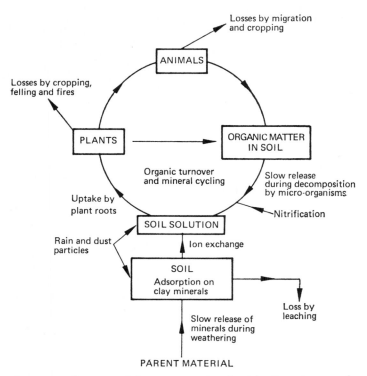

Fig. 18.1. Summary diagram of the processes involved in the release and cycling of mineral nutrients in the soil. The mineral content of the organic matter, the amount of organic matter and its rate of decomposition are important in maintenance of soil fertility. If release from parent material and from decomposition is too slow, fertility declines; if too rapid, mineral nutrients may be lost by leaching.

soils which occur over much of Britain, weathering may supply considerable quantities of the major elements required by plants, as well as trace elements and sometimes toxic minerals, the proportions depending on the nature of the rock. Other sources of mineral nutrients, as previously noted, may be flood-water, windblown dust and rain and material brought in by animals. For example, until the building of Lake Nasser the fertility of the fields of lower Egypt depended upon the yearly flooding by the silt-laden water of the Nile.

One of the essential elements for plant growth which is not supplied directly by mineralization is nitrogen. Here, as we have seen, the activities of nitrifying bacteria in converting ammonium salts to nitrates are important; nitrogen is also being continually added to the soil by the fixation of atmospheric nitrogen. This may be carried out by free-living bacteria or blue-green algae or by bacteria living symbiotically within a plant and indeed the almost universal occurrence of the bacterium *Rhizobium* in nodules on the roots of plants belonging to the pea family, the Legumin-osae, is the reason for their long-standing importance in agriculture. But rain, too, may add appreciable amounts of nitrogen to the ecosystem, especially in equatorial regions where thunderstorms are of nightly occurrence.

The prolonged movement of water downwards in the soil after rainfall carries with it dissolved material, so that, as already noted, the soil gradually becomes leached. Under semi-arid conditions, where the total rainfall is less than the potential evaporation, there will be a net upward movement of water in the soil and a crust of salts may form on the surface. However, in Britain and other humid regions the annual precipitation always exceeds the annual potential evaporation. There is, therefore, a continuing loss of nutrients to deep drainage waters and, ultimately, the ocean.

When plants are growing rapidly they absorb inorganic nutrients through their root systems. These are incorporated into their tissue and eventually returned to the soil in leaf- and fruit-fall or the death of the plant. But at times of the year when plants are not growing actively the rate of absorp-tion is reduced or may cease, as in the bluebell during the late summer when only the bulb remains, below ground and in a dormant condition. Most British plants are dormant during part of the winter, though we must remember that the root system of a deciduous tree, for instance, may be actively growing at a time when there are no leaves on the branches. The spring, when temperatures rise and active growth becomes more widespread, is also the time when the growth and activity of the soil animals, fungi and bacteria, which break down the litter of the previous autumn, also increase. Consequently the period of the most rapid release of nutrients is often the period of their most rapid uptake.

Besides the absorption of nutrients from the soil, we now know that plants can readily absorb—and release—nutrients through the surface of their leaves when rain or dew falls. The loss from leaves may be not only inorganic ions but also soluble organic compounds such as sugars and amino-acids. Although these are lost to the plant, they probably serve to encourage the growth of the soil organisms on which the plant depends for the maintenance of a good soil structure. Where the vegetation is layered the drops which fall on undergrowth may themselves have been enriched by nutrients leached out of the leaves of the canopy and these nutrients may in turn be reabsorbed by the leaves of the undergrowth. The propor-tion in which various ions occur in the washings from leaves may differ between species as well as varying with the conditions under which the plant has grown, which affect its total nutrient content. Plants on lime-rich soils tend to have more calcium in their tissues than others of the same species growing on lime-free soils, for example.

Both in loss and absorption, individual ions behave independently.

Some, such as nitrate and potassium ions, are very mobile both in the soil and in the leaf and are readily leached out by rainfall. We have seen that nitrate is continually added to the ecosystem by direct fixation of nitrogen from the atmosphere; it tends to be lost not only by the activity of soil organisms which reduce it to gaseous nitrogen but also in drainage water. Potassium is also very readily lost except near the sea, from which salt is carried inland and deposited in the rain. Leached soils quickly become poor in potash. Apart from the losses through the leaves, the flow of rain-water down stems and trunks dissolves and removes further quantities of material.

Actively growing plants, then, are continually absorbing inorganic nutrients and returning them again to the soil in leaf washings, bark- or stem-flow and the shedding of plant material above and below ground, from which they are released by the organisms of decay. This process is known as *nutrient cycling*. At any one time, the quantity of a particular element in an ecosystem can be apportioned between the soil, the vegetation and the animals. In practice it is not possible to distinguish between the amount which is incorporated in the large number of minute soil organisms and the non-living material of the soil: all is regarded as contained within the soil. In a natural ecosystem the amount contained in the larger animals is usually very small in comparison with that contained in the soil and vegetation.

Recycling of a mineral element may be exemplified with reference to phosphorus, an essential element to all living organisms, but one often in short supply. During weathering, phosphate is slowly released into the soil, and plants take up phosphate from the soil solution. The phosphates within plants may be transferred to herbivores which eat the plant material, and subsequently the phosphates may be passed to carnivores, and from them to other ('top') carnivores. Material from the various biological components of this food chain of the ecosystem—plants→herbivores→carnivores→top carnivores—ultimately reaches, by the death of the organisms, or via their waste products, the decomposers (fungi, bacteria and other micro-organisms). Consequently phosphate from all these feeding or trophic levels is ultimately released into the soil solution (sometimes after entering a series of temporary storage stages in the micro-organisms concerned), and so becomes available for uptake into plants once more (see Fig. 18.2).

In seasonal climates the ratio in which the total nutrients are partitioned between the soil and the vegetation varies from one time of the year to another, but in a stable ecosystem the amounts in the soil–vegetation complex as a whole remain almost constant. These total amounts vary from one ecosystem to another, however, and are partitioned in very different proportions in different communities. But there are also important differences between species. The supply to the surface of the soil of bases which have been absorbed in the deeper layers helps to counter-act the process of leaching and stratification characteristic under our climate. Birch has long been known to be particularly effective in this respect. Dimbleby (1952) has shown that on the Yorkshire moors invasion of heather moor by birch can lead to the breakdown of podsol and the regeneration of brown earth.

Besides the cyclical processes involving mineral nutrients within eco-systems, there may be interchange of material between ecosystems. Water

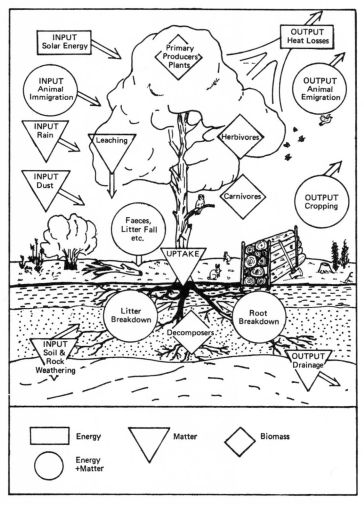

Fig. 18.2. Diagrammatic representation of the major processes involving energy and nutrient flow in a woodland ecosystem. (Adapted from J. D. Ovington, 1962, courtesy of Academic Press, London.)

containing dissolved mineral substances may drain from a terrestrial ecosystem, so depleting it, to a lake or marine ecosystem, which is thus enriched. In a woodland, timber may be felled and removed from the site, and in other communities plants cropped or animals culled. These activities and also the natural migration of organisms may all result in interchange.

ENERGY FLOW AND PRODUCTIVITY

The quantity of living plant and animal material in a given area of natural vegetation is termed the *biomass*. It can be measured in fresh weight, but more usually in terms of dry weight. One may speak of *plant biomass* and *animal biomass*. Not only does the total biomass differ greatly between biomes but the ratio of plant and animal material may vary considerably. In a deciduous oak or beechwood the total volume and weight of plant material—trees, undergrowth and ground layers—is much larger than in a meadow and in tropical rain forest it is even greater still. The animal biomass, however, is often much higher in proportion in grassland and especially so in the intensively grazed leys of the farmer. Very broadly, in undisturbed situations, the more favourable the conditions for the growth of terrestrial vegetation the greater the plant biomass.

Where the plant biomass is large the total nutrient content is also large; indeed it may contain the greater part of the nitrogen, magnesium, potassium, phosphorus, and other elements present in the soil–vegetation complex. This appears to be the case in tropical rain forests. Their magnificent growth led early European settlers to suppose the soil to be outstandingly fertile, yet after the trees were felled and cleared it was found to be of low fertility, acidic and poor in humus and nutrients. The nutrients had been removed in the act of clearing the forest. At the opposite extreme there are the deep peats of the East Anglian fenland. Considerable quantities of nutrients may be stored in the incompletely decomposed plant material of the peat. The draining and ploughing of the fens have so increased their aeration that the growth of soil micro-organisms has been stimulated, leading to the further oxidative breakdown of organic material and the release of a continuing supply of plant nutrients. But the peat is wasting away rapidly and will ultimately be lost. Under natural conditions the water retained in the peat maintained anaerobic conditions in which further breakdown did not take place.

The accumulation of humus and other more bulky organic material such as organic mud or peat depends upon a slowing down or cessation of decay. In the case of the beechwoods of southern England, beech litter decomposes to give a mull humus when growing on chalk and an acidic, superficial, mor layer on the base-poor clay-with-flints of the Chilterns. A neutral or base-rich soil favours mineralization. Waterlogging hinders it, since air is excluded. Moreover in cold climates the rate of decay is limited by temperature. Thus in general the accumulation of organic deposits tends to occur where conditions are also to some extent unfavourable for the growth of terrestrial vegetation and by the removal of nutrients from active cycling further reduces the potential rate and amount of growth of the vegetation.

As described in Chapter 4 primary succession involves a development of soil and in particular an increase in the humus content. The plant biomass is low during the early stages of succession but increases towards the climax which normally has the greatest biomass. This increase in biomass is paralleled by an increase in nutrients which become partitioned between the soil and vegetation. The total amounts of the various elements in the soil–vegetation complex, and especially nitrogen, rapidly increase at the beginning but leaching may subsequently lower the content of some, such as sodium and calcium. In a secondary succession, such as follows the colonization by weeds of abandoned ploughland, a nutrient-rich soil already exists, so that succession involves a rapid partitioning of these elements.

Plant biomass can be viewed not only as a reservoir of nutrients but as a store of chemical energy, derived in the first instance from the energy of the sun during photosynthesis. The efficiency with which plants can utilize solar energy is very small, about half of the total solar radiation received being photosynthetically active, and on average little more than 1 per cent of the energy received being fixed in the form of carbohydrates; yet it is on this very small percentage conversion that all the biological world depends. When plant material is eaten by herbivorous animals or broken down by fungal or bacterial activity a part of it is rebuilt into the structure of the animal or fungus. Some of it is oxidized in respiration, the energy thus released being used in metabolic processes, converted into potential energy or dissipated as heat. In ecological systems, energy is transferred from organism to organism (for discussion of ecological energetics see Phillipson 1966), and as energy passes through the ecosystem (Figs. 18.2, 18.3) it is steadily depleted. Energy transfer from one organism to another is always less than 100 per cent efficient because the predator needs to expend energy (which is dissipated as heat) in order to consume the prey and, in addition, assimilation of ingested food is incomplete. The total amount of energy therefore decreases at each successive trophic level. When the animal or fungus dies its tissue is in turn digested and decomposed by predators or scavengers or by decay organisms, to which it provides food for growth and the energy for their metabolism. Ultimately all the energy originally absorbed from sunlight and converted into chemical energy is dissipated as heat and lost by re-radiation into the atmosphere.

As plant biomass varies between communities so the reserves of chemical energy also vary, forests having large reserves and semi-deserts small ones.

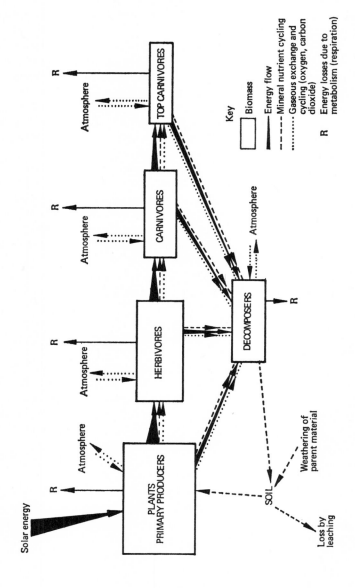

Fig. 18.3. Fundamental ecosystem dynamics. The diagram may be used to trace the flow of energy through the ecosystem, and the cycling of mineral nutrients and of carbon dioxide and oxygen.

These reserves are added to by photosynthesis and consolidated by growth. As material is constantly being lost the total remains approximately the same on a long-term basis though seasonal variation in the balance between growth and decay may be very marked. But the rate of accumulation of energy varies greatly and independently of biomass.

If we wish to know this rate of accumulation in a community it is more convenient to find it by estimating the amount of carbon fixed, that is, absorbed from the atmosphere and converted into organic material. This is done by measuring changes in the dry weight, which consists principally of carbon-containing compounds, and calculating the total energy content from samples of which the energy released in total combustion has been measured. The change in total dry weight of the plants in an ecosystem over a period of time represents the increase resulting from photosynthesis minus the loss from respiration (apparent assimilation), which is termed the *net primary production*, less the losses due to herbivores, death and shedding of parts. The net primary production in a natural ecosystem must therefore be calculated, not directly measured, and this involves estimating the amounts lost in various ways. The *rate* of this production is the *net primary productivity*. The *net secondary production* of a biome is the production of material of non-photosynthetic, heterotrophic organisms which feed directly or ultimately on the net primary production of plants. Plant productivity varies with the extent to which conditions of growth are suitable but it also varies with the species and in complex communities apparently also with interrelations between species.

It will be seen that in an undisturbed climax community, on a long-term basis, the net primary production is equalled by loss due to herbivores and death, since the biomass remains constant. But during succession there is a positive gain, at least in the early stages.

Man harvests a portion of the production of most ecosystems in the world either directly or through his domestic animals. In forests managed for timber production, a proportion of the plant biomass is removed yearly and this is usually calculated as being approximately equal to the annual increase in biomass, that is to say rather less than the net annual primary production. In other words, on a selection system of management, the timber of the trees chosen for felling balances the increment (Ovington 1965). The aim of the forester is to obtain the maximum sustained yield— the maximum yield which does not damage the forest to the extent that its productivity declines. This is also the aim of the grazier. On this basis the equivalent of the plant material which is extracted or grazed would otherwise die and decay. Overall the plant biomass is reduced but the productivity is not, indeed it may even increase.

In forestry, as in the cropping of arable land, we remove not only fixed carbon in the form of timber but also mineral nutrients and unless the rate of supply of these, from such processes as the weathering of rock particles and the fixation of nitrogen, equals the rate of their removal, there will be a continuing loss from the ecosystem. This will eventually reduce the productivity of the vegetation and may result in the replacement of the original communities by others more able to survive in an impoverished soil. But if we crop the secondary production, the animals, we may also run the same danger. One of the factors contributing to the spread of moorland and acidic grassland in hill country has been the removal of calcium and phosphorus in the bones of generations of sheep. Roberts (1959) has suggested that in North Wales the modern practice of marketing yearling lamb rather than mutton has increased the rate of nutrient loss. This feature, associated with the results of changes in other old-established practices of sward management, and especially the later return of stock to upland pastures leading to reduced grazing early in the season, has fostered the spread, for example, of the unpalatable mat grass (*Nardus stricta*).

Productivity depends upon the climate, the soil and the nature of the vegetation. The single most important factor is the supply of radiation but this can be utilized only where the temperatures are suitably high, there are sufficient available nutrients and, above all, there is adequate water. Solar radiation is greatest in tropical regions with clear, cloudless skies, but these are also arid. When they can be irrigated these semi-deserts are among the most productive regions of the world as, for example, in the Central Valley of California.

Now that there is a world shortage of land it is clearly of practical and economic importance as well as of ecological significance to discover what types of vegetation have the highest productivity and under what conditions, since ultimately all life depends on vegetation. This has been one of the principal concerns of the International Biological Programme in the 1970s.

Part V

Simple Ecological Work in the Field

Chapter 19

Ecological Studies

Ecology is of great educational value in schools and, although opportunities for such work vary widely in different schools, much can be accomplished in a school garden and by visits to the countryside.

Ecology in a wide sense is a way of looking at plants, and should never be divorced from the study of the plants themselves, of their structure, development and functions. It is appropriate to consider the relative importance of the various characters shown by plants from the ecological point of view, and in regard to their conditions of existence in the field. The ecological significance of many features of plants often then becomes clear.

Because of these considerations, close observations of the structure and of the performance of plants in the field is of first importance and the more time that can be put into such studies the better. Ecology may rightly be considered nature study *par excellence*.

ECOLOGICAL OBSERVATIONS AND INVESTIGATIONS

A good starting point is the germination of the seed and the structure and growth of seedlings, especially as the establishment of seedlings is often a critical phase in the ecology of plants. Seedlings of different plants can be grown in pots and boxes, and also in a garden if one is available. By simple experiments in which water supply and temperature may be varied, much can be learnt about the necessary conditions for germination and subsequent growth. Simply by maintaining the water table at different levels, information can be readily obtained about the effects of water supply on seedling establishment; for example, in the studies by Lazenby (1955) on the soft rush (*Juncus effusus*) it was shown that no establishment took place if the water table was as low as 8 inches (20 cm) below the surface, and

213

that the most favourable conditions for seedling establishment were when the water table was at the surface. The wide differences in structure and behaviour of seeds and seedlings of different kinds of plants, yet basic general plan common to all, are evident from studies of this kind. Also much may be learnt about the role of soil conditions, if a range of soils is used in experiments on germination and growth. Furthermore, something may be learnt about competition, for seeds of only one kind of plant may be grown in some containers or in a part of a garden, while in other instances seeds of two or more kinds of plants may be sown together and consequently their seedlings develop in competition with one another.

Such studies on soil, seeds and seedlings can profitably be done during the winter, especially if the germination of seeds is begun indoors in February and March, and seedling growth later continued out-of-doors at a time when seedlings are springing up vigorously everywhere. In the autumn a study can be made of fruits and seeds, of the numbers in which they are formed, of seed viability, and of their means and distance of natural dispersal. In woodland such details could be observed for the commoner British trees including oak, beech, ash, birch and pine, and the work supplemented by observations on leaf-fall, and of the winter conditions of the trees, their shape and the form and colour of the twigs, winter buds and bark. In the late spring and summer, the leafing of trees can be examined; the shade which they cast and the shade which is tolerated by seedlings and young trees can be compared. Species of the ground flora can be studied, different types of life form (see Chapter 5) being examined and compared. The flower structure and flowering behaviour of trees, shrubs and herbs form a further study. During the summer simple experiments on growth, transpiration and photosynthesis can also be conducted.

Invaluable first-hand acquaintance of many features of ecological interest can be gained from a school garden. Included here are characters of the soil, and the effects of cultivation; the growth of plants of different life forms from seed, rootstocks, corms, bulbs and other organs of vegetative propagation; the effects of weather conditions on different plants; and studies on weeds. The comparison between the relatively harmless and easily removed annual weeds such as shepherd's purse (*Capsella bursa-pastoris*) and groundsel (*Senecio vulgaris*) with persistent perennial weeds with deeply-rooted, quickly growing rhizomes such as couch-grass (*Agropyron repens*) and bindweed (*Convolvulus arvensis*) is a good demonstration of different life forms and contrasted ecological relations. The garden, besides having well-tended beds of flowers and vegetables, can be viewed as an outdoor laboratory for observing the growth of a wide range of plants and the conditions under which they succeed or fail. Here repeated observations of the growth of plants can be made, not only of the above-ground shoot system, but also of the underground parts. Seedlings and older plants can be dug up, and the depth at which the underground parts grow explored as well as their nature and development.

Such studies can be profitably followed up by observations made in the open country. For example, sets of seedlings seen on waysides, hedgebanks, the edges of woods, and in arable fields can be marked (by small stakes) and observed week by week and their development noted. Some will die and disappear from causes which are not obvious, whereas some will be smothered by the growth of neighbouring plants, or will compete with one another for light, grow long and spindly and finally die. Some may be attacked and damaged or destroyed by insects or parasitic fungi; others may be trampled down or eaten by browsing or nibbling animals. Yet others may suffer from drought, because of lack of soil water. Some, however, will grow into adults and their weekly progress can be recorded. It is informative to dig up specimens at various stages, wash the roots carefully free from soil and make accurate drawings to scale. In this way, as in other instances in ecological study, classroom work makes a valuable supplement to observations in the field, garden and laboratory.

A further valuable approach is to choose a limited range of plants for the detailed study of as many of their features as possible; mention has already been made of such autecological work. It is wise to choose fairly common plants for this study, if possible characteristic of different communities, e.g. a woodland plant such as dog's mercury (*Mercurialis perennis*) or common violet (*Viola riviniana*), a heath plant such as ling (*Calluna vulgaris*), a wayside plant such as silverweed (*Potentilla anserina*) or one of the plantains (*Plantago* spp.), one or two pasture plants including a dominant grass, and so on. To these may be added two or three annuals, preferably those whose development has been followed from the seedling. In an agricultural region, plants may be chosen from among arable weeds, and also dominant or characteristic plants from any areas of comparatively natural vegetation within easy reach.

Observations should be made on as many points as possible about the plants chosen. Included here are life form and mode of perennation; depth at which the roots are mainly developed (effective rooting depth); general structure of the shoot system and means of vegetative propagation, if any; and course of development of the plant, time of leaf and of flower production, duration of the flower, flower structure and mode of pollination. Direct observations of insects visiting flowers and their behaviour will often indicate the method of cross-pollination (the simple procedure of tying a muslin bag over flowers will show whether large insects are necessary). The amount of seed, if any, which is set can be examined, and its viability tested. The partitioning of the activity of the plant in terms, for example, of roots, leaves, stems, flowers and seeds at different stages of its development is of considerable ecological interest (Harper 1967), and can be assessed by fairly simple measurements. In particular, the importance of spread by vegetative means relative to that by seeds can be estimated, and the relative energy expended into vegetative propagation and seed production determined; such studies bear on what may be called the 'reproductive strategy' of the plant. The performance of the plant in its habitat can also be assessed, and its relations with other plants (i.e. studies can be made of its competitive ability). Observations over a wide area are desirable, as these will probably throw light on the various soil conditions in which the plant can succeed, and the different communities in which the plant can occur.

Chapter 20

The Development of Ecological Work

Ecological studies can best be developed by a combination of regular school classes and field observations reinforced by the voluntary work of students out-of-doors and by the activities of Field Clubs and Natural History Societies. Work may be undertaken either on intensive or extensive lines. Advantages of intensive work are that it does not require a wide knowledge of the flora, it can be confined to small plots of ground and it directs attention to points of detail; a limited range of plants can be investigated thoroughly. Extensive work necessitates free access to the countryside and can be most profitably undertaken when some knowledge of the flora has been gained.

QUADRAT STUDIES

For intensive work the quadrat (see Chapter 10) is invaluable, and particularly the permanent quadrat which is listed at intervals. It may be possible to lay out a range of quadrats in a school garden, especially if there are marked differences in the soil, and if some parts are wetter or more shaded than others. Access to each quadrat is required in all states of the soil; also as the observer needs to get very close to every part of the surface it is necessary to kneel or lie beside it (polythene sheets give good protection from the soil).

Garden quadrats may usefully be laid out on bare soil in the winter and charted as soon as the first crop of seedlings appears and thereafter at appropriate intervals. Although seedlings will be seen to have different types of cotyledons, it will often not be possible to identify many of the plants until they reach later stages; until then, suitable symbols may be used to denote unknown seedlings. Careful attention must be given to the scale on which the quadrat is charted, and this must be noted for every quadrat.

As the seedlings grow, and the distinctive characteristics of the different plants become evident, the species to which they belong will gradually be identified, and in this way an extensive knowledge of garden weeds and of their rate and mode of development can be acquired (the Ministry of Agriculture, Fisheries and Food Bulletin (1961) is a useful aid in the identification of the seedlings of common weeds). Comparison of successive charts (which must be carefully dated) of the different quadrats from year to year will furnish much instructive and interesting information. Different quadrats will almost certainly show somewhat different plant populations, although some species will be in common. The differences will depend, for example, on what seeds were initially present, features of the habitat, such as soil conditions, water supply and shade, and the proximity of seed parents. These causes of differences furnish useful points for further study. Once it has been established what plants are in the quadrat, and in what numbers, the question arises as to why these particular plants are present, and in these particular numbers and with a particular type of distribution. As a result of careful observation, other questions will arise, and attempts to find answers to these will lead to a greater understanding of the situation.

Besides the first colonization, there will be a succession of plants from year to year in the quadrats; some species remain for a long time, whereas others are transitory. Information bearing on this can be obtained by records over the years; it is clearly desirable that permanent quadrats should remain undisturbed for as long as possible, and that no plants are pulled up but are left to rot where they die.

Much detailed knowledge of the behaviour of plants may be acquired from quadrat studies even in a single season. Features which would probably be missed entirely are noticed, and an extraordinarily close picture of growth and development is gained from the re-observation of individual plants at definite intervals. Besides furnishing information on succession, quadrats can be set out and experiments tried on them. For example, the effects of over-watering may be studied, of protection from rain, of various kinds and degrees of shading, of various kinds of manuring and fertilizer treatment, and seeds introduced in manure may be determined.

Permanent quadrats may also be used in established plant communities, but here there is not very much change from year to year and sites where there is little danger of disturbance are often hard to find. Temporary quadrats, however, are very useful for providing information on the occurrence and frequency of species. Where colonization is proceeding on bare or partially bare soil, semi-permanent quadrats are valuable in the study of succession and need not be charted so frequently as in a garden because growth and change are less rapid.

Line transects (see Chapter 10) provide particularly useful information in zoned vegetation when taken at right-angles to the direction of the zones. Several line transects can be quickly scored by a party and a very long one can be easily divided into sections to be recorded by different groups of a party.

EXTENSIVE WORK IN THE FIELD

For a study of plant communities, knowledge of the flora is essential and this is most appropriately gradually built up by the use of floras such as the *Excursion Flora of the British Isles.* In this way an acquaintance with the more common flowering plants can be gained and knowledge of rarer plants can be added from time to time when they are encountered in the various habitats investigated.

In community studies, attention may also be profitably directed towards the life forms of plants (Chapter 5), and a certain correspondence between life form and habitat will often become obvious. For example, in reed-swamps there is a preponderance of plants with tall slender shoots; in loose or soft soils many of the plants have long underground shoots; annuals are dominant in ploughland, gardens and on bare soil generally; and in pastures, turf-forming species, notably grasses, which branch freely from buds close to the soil surface, are the most important plants.

Selection of particular communities for study must depend mainly on what is readily accessible. Heath and common land are often very suitable; woodland provides a wide field of work; marsh and aquatic vegetation, while interesting, may present difficulties in respect of repeated studies; but coastal vegetation affords unrivalled opportunities.

Various ways of studying communities have been indicated in previous chapters. It is useful at an early stage to identify the plants as fully as possible and to compile a list of species as complete as possible. If there are marked differences in conditions in different parts of the habitat, or if succession is obviously proceeding, a general vegetation map, on an appropriate scale, is of value. The structure of the vegetation can be examined by means of quadrats, or, especially in zoned vegetation, also by the use of transects. Where possible, habitat factors should be considered, and desirably measured (see Chapter 14), and the effects of grazing (if any) noted. Simple experiments, arising from observed features, can often be undertaken in this connection and provide revealing results. Other studies, e.g. of the effect of fires, and of insect attack, of the rate of spread of plants, and of the extent of seed production and seedling establishment, can be made as appropriate in the particular habitats investigated.

Man-made habitats often repay close study; the effects of trampling and of pollution can frequently be observed and quantitative assessments of these effects made. Cleared sites are often ideal for the study of succession, while the colonization of walls and ditches presents many features of interest. Disused railway lines may support a diverse flora, and a railway cutting which runs east to west affords a good opportunity to study the influence of aspect; the vegetation of the north-facing and of the south-facing bank may differ appreciably both in the plants present and in their seasonal behaviour and growth.

In all these different studies the importance of making quantitative observations will become clear. This will necessitate sampling procedures (see Chapter 11) and the scale of the sampling operation will have to be decided. It is usually impossible to count or measure everything (and if this is attempted damage may sometimes be caused), so a random sample is selected.

Besides studies on flowering plants, it is important that the ecologist should not neglect the ferns and the non-vascular 'lower' plants—mosses, liverworts, lichens, algae and fungi. These plants are often important in communities dominated by higher plants, and may form distinct minor communities of their own. Some of them play a key part in succession, especially in the early stages of colonization where lichens and mosses are frequently pioneers. Fungi play a particularly significant ecological role in the soil and in the decomposition of dead plant material. Many of the common species of lower plants can be identified by reference to standard handbooks on the groups (see list, p. 222), and field observations may suggest their role in the ecosystem. Such studies help to reflect the comprehensive nature of plant ecology and promote the development of first-hand knowledge of important elements of the environment.

Appendix

Sources of supply of maps and certain equipment useful in ecology

SUPPLIERS OF MAPS

Ordnance Survey maps

These are available from Edward Stanford Ltd, 12–14 Long Acre, London, W.C.2, Thomas Nelson & Sons Ltd, 18 Dalkeith Road, Edinburgh 6, local Ordnance Survey agents and from most booksellers.

Information about the availability of Ordnance Survey publications is obtainable from The Director General, Ordnance Survey, Romsey Road, Maybush, Southampton, SO9 4DH.

Soil Survey maps

Soil Survey maps of the Soil Survey of Great Britain are obtainable in the same way as other Ordnance Survey publications.

Bulletins, Memoirs, Soil Survey Records and the *Soil Survey Field Handbook* (1960) are available from The Soil Survey of England and Wales, Rothamsted Experimental Station, Harpenden, Herts. Information concerning Scotland is obtainable from The Soil Survey of Scotland, The Macaulay Institute for Soil Research, Craigiebuckler, Aberdeen.

Geological Survey maps

Obtainable as other Ordnance Survey publications. Information is available from The Director, The Institute of Geological Sciences, Exhibition Road, South Kensington, London, S.W.7.

Land-Use Survey maps

Enquiries to Miss A. M. Coleman, M.A., Department of Geography, King's College, Strand, London, W.C.2, or Edward Stanford Ltd (see above).

SUPPLIERS OF SOME EQUIPMENT USEFUL IN ECOLOGICAL INVESTIGATIONS

Light meters involving photovoltaic cells of the selenium barrier type

Evans Electroselenium Ltd, St Andrews Works, Halstead, Essex, and Megatron Ltd, 115a Fonthill Road, Finsbury Park, London, N.4.

Photoresistors (cadmium sulphide cells)

Mullard Ltd, Mullard House, Torrington Place, London, W.C.1, and Proops Bros. Ltd, 52 Tottenham Court Road, London, W.1.

General Reference Books

The following is a selection of books useful in identification.

FLORAS

Clapham, A. R., Tutin, T. G. and Warburg, E. F. (1962) *Flora of the British Isles*, 2nd edn. Cambridge University Press, London.

Clapham, A. R., Tutin, T. G. and Warburg, E. F. (1968) *Excursion Flora of the British Isles*, 2nd edn. Cambridge University Press, London.

Fitter, R. S. R. and McClintock, D. (1956) *Collins Pocket Guide to Wild Flowers.* Collins, Glasgow.

Hubbard, C. E. (1968) *Grasses*, 2nd edn. Penguin (Pelican series), London.

PLANT DISTRIBUTION

Perring, F. H. and Walters, S. M. (eds) (1962) *Atlas of the British Flora.* Nelson, London.

For local distribution, reference may be made to County Floras (now available for most counties).

LOWER PLANTS

Dixon, H. N. (1904) *The Student's Handbook of British Mosses*, 2nd edn. Sumfield, Eastbourne.

MacVicar, S. M. (1926) *The Student's Handbook of British Hepatics*, 2nd edn. Sumfield, Eastbourne.

Watson, E. V. (1968) *British Mosses and Liverworts*, 2nd edn. Cambridge University Press, London.

Alvin, K. L. and Kershaw, K. A. (1963) *The Observer's Book of Lichens.* Warne, London.

Lange, M. and Hora, F. B. (1963) *Collins Guide to Mushrooms and Toadstools.* Collins, London.

Bibliography

ADAMSON, R. S. (1912) An ecological study of a Cambridgeshire woodland. *J. Linn. Soc. Bot.* **40**, 339–87.

ADAMSON, R. S. (1918) On the relationships of some associations of the southern Pennines. *J. Ecol.* **6**, 97–109.

ANDERSON, D. J. (1961) The structure of some upland plant communities in Caernarvonshire. II. The pattern shown by *Vaccinium myrtillus* and *Calluna vulgaris*. *J. Ecol.* **49**, 731–8.

ANDERSON, M. C. (1964*a*) Studies of the woodland light climate. I. The photographic computation of light conditions. *J. Ecol.* **52**, 27–41.

ANDERSON, M. C. (1964*b*) *Ibid.* II. Seasonal variation in the light climate. *J. Ecol.* **52**, 643–63.

AVASTE, O., MOLDAU, H. and SCHIFRIN, K. S. (1962) Spectral distribution of direct solar and diffuse radiation. *Akad. Nauk. Est. SSR Inst. Phys. Astron.* Investigations in Atmospheric Physics No. 3.

BAINBRIDGE, R., EVANS, G. C. and RACKHAM, O. (1966) *Light as an Ecological Factor.* Symposium vol. 6, British Ecological Society. Blackwell Scientific Publications, Oxford.

BLACKMAN, G. E. (1935) A study by statistical methods of the distribution of species in grassland associations. *Ann. Bot., Lond.* **49**, 749–77.

BLACKMAN, G. E. and RUTTER, A. J. (1946–50) Physiological and ecological studies in the analysis of plant environment. Parts I–V. *Ann. Bot., n.s.*, **10**, 361–90 (1946); **11**, 125–58 (1947); **12**, 1–26 (1948); **13**, 453–89 (1949); **14**, 487–520 (1950).

BRAUN-BLANQUET, J. (1932) *Plant Sociology: The Study of Plant Communities* (English translation by G. D. Fuller and H. S. Conard). McGraw-Hill, New York.

BURGES, A. (1958) *Micro-organisms in the Soil.* Hutchinson, London.

BURNETT, J. H. (ed.) (1964) *The Vegetation of Scotland.* Oliver & Boyd, Edinburgh.

BURNHAM, C. P. and MACKNEY, D. (1964) Soils of Shropshire. *Field Studies*, **2**, 83–113.

CAMPBELL, R. C. (1967) *Statistics for Biologists.* Cambridge University Press, London.

CLAPHAM, A. R. (1932) The form of the observational unit in quantitative ecology. *J. Ecol.* **20**, 192–7.

CLAPHAM, A. R. (1936) Over-dispersion in grassland communities and the use of statistical methods in plant ecology. *J. Ecol.* **24**, 232–51.

CLARKE, G. M. (1969) *Statistics and Experimental Design.* Edward Arnold, London.

CLIFFORD, H. T. (1956) Seed dispersal on footwear. *Proc. bot. Soc. Br. Isl.* **2**, 129–31.

CONWAY, V. M. (1936–8) Studies in the autecology of *Cladium mariscus* R. Br. Parts I–V. *New Phytol.* **35**, 177–204 (1936); **35**, 359–80 (1936); **36**, 64–96 (1937); **37**, 254–78 (1938); **37**, 312–28 (1938).

DARLINGTON, A. (1969) *Ecology of Refuse Tips.* Heinemann Educational Books, London.

DIMBLEBY, G. W. (1952) Soil regeneration on the north-east Yorkshire moors. *J. Ecol.* **40**, 331–41.

DOWDESWELL, W. H. and HUMBY, S. R. (1953) A photovoltaic light meter for school use. *Sch. Sci. Rev.* **35** (125), 64–70.

EVANS, G. C. (1956) An area survey method of investigating the distribution of light intensity in woodlands, with particular reference to sunflecks. *J. Ecol.* **44**, 391–428.

EVANS, G. C. and COOMBE, D. E. (1959) Hemispherical and woodland canopy photography and the light climate. *J. Ecol.* **47**, 103–13.

223

FARROW, E. P. (1916) On the ecology of the vegetation of Breckland. II. Factors relating to the relative distributions of *Calluna*-heath and grass-heath in Breckland. *J. Ecol.* **4**, 57–64.

FARROW, E. P. (1917*a*) *Ibid.* III. General effects of rabbits on the vegetation. *J. Ecol.* **5**, 1–18.

FARROW, E. P. (1917*b*) *Ibid.* IV. Experiments mainly relating to the available water supply. *J. Ecol.* **5**, 104–13.

FARROW, E. P. (1917*c*) *Ibid.* V. Observations relating to competition between plants. *J. Ecol.* **5**, 155–72.

GATES, D. M. (1962) *Energy Exchange in the Biosphere*. Harper & Row Biological Monographs, New York.

GILLHAM, M. E. (1970) Seed dispersal by birds. In *The Flora of a Changing Britain* (ed. F. H. Perring), pp. 90–8, Botanical Society of the British Isles; Classey, Hampton, Middlesex.

GIMINGHAM, C. H. (1949) The effects of grazing on the balance between *Erica cinerea* L. and *Calluna vulgaris* (L.) Hull in upland heath, and their morphological responses. *J. Ecol.* **37**, 100–19.

GIMINGHAM, C. H. (1960) Biological Flora of the British Isles: *Calluna vulgaris* (L.) Hull. *J. Ecol.* **48**, 455–83.

GODWIN, H. (1956) *The History of the British Flora*. Cambridge University Press, London.

GREGORY, R. P. G. and BRADSHAW, A. D. (1965) Heavy metal tolerance in populations of *Agrostis tenuis* Sibth. and other grasses. *New Phytol.* **64**, 131–43.

GREIG-SMITH, P. (1964) *Quantitative Plant Ecology*, 2nd edn. Butterworths, London.

GRIME, J. P. and HUTCHINSON, T. C. (1967) The incidence of lime-chlorosis in the natural vegetation of England. *J. Ecol.* **55**, 557–66.

GRUBB, P. J., GREEN, H. E. and MERRIFIELD, R. C. J. (1969) The ecology of chalk heath: its relevance to the calcicole–calcifuge and soil acidification problems. *J. Ecol.* **57**, 175–212.

HANDLEY, W. R. C. (1963) Mycorrhizal associations and *Calluna* heathland afforestation. *Forestry Commission Bulletin* **36**, H.M.S.O., London.

HARPER, J. L. (1967) A Darwinian approach to plant ecology. *J. Ecol.* **55**, 247–70.

HESSE, P. R. (1971) *A Textbook of Soil Chemical Analysis*. John Murray, London.

HOLLAND, P. G. (1972) The pattern of species density of old stone walls in western Ireland. *J. Ecol.* **60**, 799–805.

HOPE-SIMPSON, J. F. (1938) A chalk flora on the Lower Greensand: its use in interpreting the calcicole habit. *J. Ecol.* **26**, 218–35.

HOPE-SIMPSON, J. F. (1940) On the errors in the ordinary use of subjective frequency estimations in grassland. *J. Ecol.* **28**, 193–209.

HOPKINS, B. (1954) A new method for determining the type of distribution of plant individuals. *Ann. Bot., n.s.*, **18**, 213–27.

HOPKINS, B. (1957) Pattern in the plant community. *J. Ecol.* **45**, 451–63.

HUTCHINSON, T. C. (1967) Coralloid root systems in plants showing lime-induced chlorosis. *Nature, Lond.* **214**, 943–5.

INGRAM, M. (1958) The ecology of the Cairngorms. IV. The *Juncus* zone: *Juncus trifidus* communities. *J. Ecol.* **46**, 707–37.

JACKSON, J. E. and SLATER, C. H. W. (1967) An integrating photometer for outdoor use particularly in trees. *J. appl. Ecol.* **4**, 421–4.

JACKSON, M. L. (1962) *Soil Chemical Analysis*. Constable, London.

JACKSON, R. M. and RAW, F. (1966) *Life in the Soil*. Studies in Biology, no. 2. Edward Arnold, London.

JEFFERIES, R. L. and WILLIS, A. J. (1964) Studies on the calcicole–calcifuge habit. II. The influence of calcium on the growth and establishment of four species in soil and sand cultures. *J. Ecol.* **52**, 691–707.

JEFFREYS, H. (1917) On the vegetation of four Durham Coal-Measure Fells. III. On water-supply as an ecological factor. *J. Ecol.* **5**, 129–40.

JONES, H. (1955) A device for obtaining photographs from an appreciable height above the ground. *J. Ecol.* **43**, 72–3.

KERSHAW, K. A. (1964) *Quantitative and Dynamic Ecology*. Edward Arnold, London.

KNIGHT, G. H. (1964) Some factors affecting the distribution of *Endymion non-scriptus* (L.) Garcke in Warwickshire woods. *J. Ecol.* **52**, 405–21.

LAZENBY, A. (1955) Germination and establishment of *Juncus effusus* L. II. The interaction effects of moisture and competition. *J. Ecol.* **43**, 595–605.

LEE, J. A. and WOOLHOUSE, H. W. (1969) A comparative study of bicarbonate inhibition of root growth in calcicole and calcifuge grasses. *New Phytol.* **68**, 1–11.

McVEAN, D. N. (1953) Biological Flora of the British Isles: *Alnus* Mill. *J. Ecol.* **41**, 447–66.

McVEAN, D. N. (1955–9) Ecology of *Alnus glutinosa* (L.) Gaertn. *J. Ecol.* **43**, 46–60; **43**, 61–71; **44**, 195–218; **44**, 219–25; **44**, 321–30; **44**, 331–3; **47**, 615–18.

McVEAN, D. N. (1963) Ecology of Scots pine in the Scottish Highlands. *J. Ecol.* **51**, 671–86.

McVEAN, D. N. and RATCLIFFE, D. A. (1962) *Plant Communities of the Scottish Highlands*. Monographs of the Nature Conservancy, no. 1. H.M.S.O., London.

MINISTRY OF AGRICULTURE, FISHERIES AND FOOD (1961) Identification of seedlings of common weeds. *Bulletin* No. 179. H.M.S.O., London.

MONTEITH, J. L. (1973) *Principles of Environmental Physics*. Edward Arnold, London.

MUNSELL, A. H. (1954) *A Color Notation*, 10th edn. Munsell Color Co., Baltimore, Md.

NEWMAN, E. I. (1963) Factors controlling the germination date of winter annuals. *J. Ecol.* **51**, 625–38.

OVINGTON, J. D. (1962) Quantitative ecology and the woodland ecosystem concept. In *Advances in Ecological Research*, vol. 1 (ed. J. B. Cragg), pp. 103–92. Academic Press, London.

OVINGTON, J. D. (1965) *Woodlands*. English Universities Press, London.

PEARSALL, W. H. (1917, 1918) The aquatic and marsh vegetation of Esthwaite Water. *J. Ecol.* **5**, 180–202; **6**, 53–74.

PEARSALL, W. H. (1938) The soil complex in relation to plant communities. II. Characteristic woodland soils. *J. Ecol.* **26**, 194–209. III. Moorlands and bogs. *J. Ecol.* **26**, 298–315.

PEARSALL, W. H. (1971) *Mountains and Moorlands* (revised edn). The New Naturalist, Collins, London.

PHILLIPS, E. A. (1959) *Methods of Vegetation Study*. Holt-Dryden, U.S.A.

PHILLIPSON, J. (1966) *Ecological Energetics*. Studies in Biology, no. 1. Edward Arnold, London.

PIGOTT, C. D. (1968) Biological Flora of the British Isles: *Cirsium acaulon* (L.) Scop. *J. Ecol.* **56**, 597–612.

PIGOTT, C. D. and TAYLOR, K. (1964) The distribution of some woodland herbs in relation to the supply of nitrogen and phosphorus in the soil. *J. Ecol.* **52** (suppl.), 175–85.

POORE, M. E. D. (1955a) The use of phytosociological methods in ecological investigations. I. The Braun-Blanquet system. *J. Ecol.* **43**, 226–44.

POORE, M. E. D. (1955b) *Ibid.* II. Practical issues involved in an attempt to apply the Braun-Blanquet system. *J. Ecol.* **43**, 245–69.

POWELL, M. C. and HEATH, O. V. S. (1964) A simple and inexpensive integrating photometer. *J. exp. Bot.* **15**, 187–91.

RAUNKIAER, C. (1934) *The Life Forms of Plants and Statistical Plant Geography.* Clarendon Press, Oxford.

RICHARDS, P. W. (1939) Ecological studies on the rain forest of Southern Nigeria. I. The structure and floristic composition of the primary forest. *J. Ecol.* **27**, 1–61.

RIDLEY, H. N. (1930) *The Dispersal of Plants Throughout the World.* Reeve, Ashford, Kent.

ROBERTS, R. ALUN (1959) Ecology of human occupation and land use in Snowdonia. *J. Ecol.* **47**, 317–23.

RORISON, I. H. (1960) The calcicole–calcifuge problem. II. The effects of mineral nutrition on seedling growth in solution culture. *J. Ecol.* **48**, 679–88.

RORISON, I. H. (1967) A seedling bioassay on some soils in the Sheffield area. *J. Ecol.* **55**, 725–41.

SALISBURY, E. J. (1916) The oak-hornbeam woods of Hertfordshire. Parts I and II. *J. Ecol.* **4**, 83–117.

SALISBURY, E. J. (1918) *Ibid.* Parts III and IV. *J. Ecol.* **6**, 14–52.

SALISBURY, E. J. (1942) *The Reproductive Capacity of Plants.* G. Bell & Sons, London.

SALISBURY, E. J. (1952) *Downs and Dunes: Their Plant Life and its Environment.* G. Bell & Sons, London.

SALISBURY, E. J. (1964) *Weeds and Aliens*, 2nd edn. The New Naturalist, Collins, London.

SHIMWELL, D. W. (1971) *The Description and Classification of Vegetation.* Sidgwick & Jackson, London.

SMITH, A. D. (1944) A study of the reliability of range vegetation estimates. *Ecology*, **25**, 441–8.

SNAYDON, R. W. and BRADSHAW, A. D. (1961) Differential response to calcium within the species *Festuca ovina* L. *New Phytol.* **60**, 219–34.

SOIL SURVEY STAFF (1960) *Field Handbook.* Soil Survey of Great Britain.

SZEICZ, G. (1968) Measurement of radiant energy. In *The Measurement of Environmental Factors in Terrestrial Ecology* (ed. R. M. Wadsworth), pp. 109–30. Blackwell Scientific Publications, Oxford.

TANSLEY, A. G. (1917) On competition between *Galium saxatile* L. (*G. hercynicum* Weig.) and *Galium sylvestre* Poll. (*G. asperum* Schreb.) on different types of soil. *J. Ecol.* **5**, 173–9.

TANSLEY, A. G. (1922) Studies of the vegetation of the English Chalk. II. Early stages of redevelopment of woody vegetation on chalk grassland. *J. Ecol.* **10**, 168–77.

TANSLEY, A. G. (1953) *The British Islands and Their Vegetation.* Cambridge University Press, London.

TANSLEY, A. G. (1968) *Britain's Green Mantle* (2nd edn, revised by M. C. F. Proctor). Allen & Unwin, London.

THOMAS, A. S. (1960) Changes in vegetation since the advent of myxomatosis. *J. Ecol.* **48**, 287–306.

THOMAS, A. S. (1963) Further changes in vegetation since the advent of myxomatosis. *J. Ecol.* **51**, 151–86.

TREPP, W. (1950) Ein Beitrag zu Bonitier ungamethoden von Alpweiden. *Schweiz. landw. Mh.* **28**, 366–71.

WARDLE, P. (1959) The regeneration of *Fraxinus excelsior* in woods with a field layer of *Mercurialis perennis*. *J. Ecol.* **47**, 483–97.

WARREN WILSON, J. (1959) Notes on wind and its effects in arctic-alpine vegetation. *J. Ecol.* **47**, 415–27.

WATT, A. S. (1919) On the causes of failure of natural regeneration in British oakwoods. *J. Ecol.* **7**, 173–203.

WATT, A. S. (1923) On the ecology of British beechwoods with special reference to their regeneration. Part I. The causes of failure of natural regeneration of the beech (*Fagus silvatica* L.). *J. Ecol.* **11**, 1–48.

WATT, A. S. (1924, 1925) *Ibid.* Part II. The development and structure of beech communities on the Sussex Downs. *J. Ecol.* **12**, 145–204; **13**, 27–73.

WATT, A. S. (1931) Preliminary observations on Scottish beechwoods. Parts I and II. *J. Ecol.* **19**, 137–57; 321–59.

WATT, A. S. (1934a) The vegetation of the Chiltern Hills, with special reference to the beechwoods and their seral relationships. Part I. *J. Ecol.* **22**, 230–70.

WATT, A. S. (1934b) *Ibid.* Part II. *J. Ecol.* **22**, 445–507.

WATT, A. S. (1947a) Pattern and process in the plant community. *J. Ecol.* **35**, 1–22.

WATT, A. S. (1947b) Contributions to the ecology of bracken (*Pteridium aquilinum*). IV. The structure of the community. *New Phytol.* **46**, 97–121.

WEAVER, J. E. (1919) *The Ecological Relations of Roots.* Publication No. 286, Carnegie Institution of Washington.

WEBSTER, J. R. (1962) The composition of wet-heath vegetation in relation to aeration of the ground-water and soil. II. Response of *Molinia coerulea* to controlled conditions of soil aeration and ground-water movement. *J. Ecol.* **50**, 639–50.

WELL, T. C. E. (1969) Botanical aspects of conservation management of chalk grasslands. *Biological Conservation,* **2**, 36–44.

WILLIAMS, W. T. and LAMBERT, J. M. (1959) Multivariate methods in plant ecology. I. Association-analysis in plant communities. *J. Ecol.* **47**, 83–101.

WILLIAMS, W. T. and LAMBERT, J. M. (1960) *Ibid.* II. The use of an electronic digital computer for association-analysis. *J. Ecol.* **48**, 689–710.

WILLIS, A. J. (1963) Braunton Burrows: the effects on the vegetation of the addition of mineral nutrients to the dune soils. *J. Ecol.* **51**, 353–74.

WILLIS, A. J. (1965) The influence of mineral nutrients on the growth of *Ammophila arenaria. J. Ecol.* **53**, 735–45.

WILLIS, A. J. and DAVIES, E. W. (1960) *Juncus subulatus* Forsk. in the British Isles. *Watsonia,* **4**, 211–17.

WILLIS, A. J., FOLKES, B. F., HOPE-SIMPSON, J. F. and YEMM, E. W. (1959) Braunton Burrows: the dune system and its vegetation. Part I. *J. Ecol.* **47**, 1–24.

WILLIS, A. J. and YEMM, E. W. (1961) Braunton Burrows: mineral nutrient status of the dune soils. *J. Ecol.* **49**, 377–90.

WOODRUFFE-PEACOCK, E. A. (1918) A fox-covert study. *J. Ecol.* **6**, 110–25.

Index

Page numbers in bold type refer to illustrations

30, 960

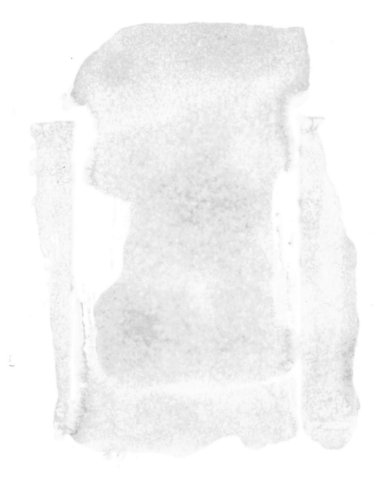